the return of the unicorns

BIOLOGY AND RESOURCE MANAGEMENT

the return of the unicorns

THE NATURAL HISTORY AND
CONSERVATION OF THE
GREATER ONE-HORNED RHINOCEROS

Eric Dinerstein

COLUMBIA UNIVERSITY PRESS   NEW YORK

Columbia University Press
Publishers Since 1893
New York   Chichester, West Sussex

Copyright © 2003 Columbia University Press

Library of Congress Cataloging-in-Publication Data
Dinerstein, Eric, 1952–
    The return of the unicorns : the natural history and conservation of the
greater one-horned rhinoceros / Eric Dinerstein.
        p. cm. — (Biology and resource management)
    Includes bibliographical references and index.
    ISBN 0-231-08450-1 (cloth : alk. paper)
    1. Indian rhinoceros. 2. Endangered species — Asia. 3. Wildlife
conservation — Nepal — Royal Chitwan National Park.   I. Title.
    QL737.U63 D56 2003
    639.97'9668 — dc21   2002034993

Columbia University Press books are printed on permanent and durable acid-free
paper.
Designed by Chang Jae Lee
Printed in the United States of America

c 10 9 8 7 6 5 4 3 2 1

This book is dedicated to Chris Wemmer, my teacher, friend, and mentor at times when I needed all three.

CONTENTS

FOREWORD

AS I SAT ON THE ELEPHANT, I looked past its bristly skull through a bower of grass arching over our trail in Chitwan. The only sound was the swish of grass against the animal's body as it glided along. We emerged abruptly into a clearing where a gray hulking creature with skin folds like armor plate faced us, its nostrils flared, its ears twitching. The curved horn on its nose looked menacing. We halted. The scene was anachronistic, antediluvian, two giants of the earth meeting as they have for eons. It was 1972, but I felt adrift in time, a bystander in a Pliocene moment. The rhino snorted with irritation, and then lumbered off into the tall grass.

Although I rejoiced in the encounter, I also had feelings of sadness and guilt. This rhino was a survivor, the product of at least 35 million years of evolution, its species reduced to perhaps no more than 1,000 individuals in several reserves in India and in the Chitwan area of Nepal. Demand for its land, the fertile floodplains, and demand for its body parts, whether for the horn that is used in traditional Chinese medicine to reduce fever or the blood taken as a tonic in India, has inexorably pushed this extraordinary creature toward extinction. India's king used tame rhinos in battles in the late fourteenth century, and the maharaja of Cooch Behar shot 207 rhinos around the turn of the nineteenth century. But some values have changed, and now rhinos have become a major tourist attraction, symbols of a vanished age.

The basic issues, the most urgent today, are to conserve the diversity of life and to maintain ecological processes, to leave future generations with options. Fortunately, Nepal and India have responded to the plight of the rhino by giving it vigorous protection, not for economic reasons but mainly because of pride in this unique natural heritage. However, detailed research on a species and its environment is essential before realistic long-term conservation plans can be made. Natural history provides the basis for understanding the requirements of a species, something that researchers often disregard in this age of remote sensing and computer modeling. Yet describing the life of an animal is not enough: a researcher today also has a moral responsibility to help the species endure. To probe an animal is for the most part enjoyable, even if it takes a capacity to endure discomfort. By contrast, conservation is an extremely difficult task because it deals more with economics and social issues than biology. A naturalist must become a politician, educator, anthropologist, and fund-raiser to achieve any success. Furthermore, a conservation project is endless, a species or habitat never completely safe, because local situations never remain static, and new problems constantly intrude. Conservation requires individuals with a special combination of talents.

I mention these matters because Chitwan is a success story, years removed from 1959 when poachers slaughtered at least 75 rhinos. In 1973 the government declared Chitwan a national park. Soon thereafter the military arrived to help protect the area, and the park was increased in size. When I visited in 1972, it had about 100 rhinos; now the population is more than 600. Since 1984, Eric Dinerstein has dedicated himself to the rhinos of Chitwan; he is the best friend they have ever had. Filled with enthusiasm and curiosity, he spent four years studying daily movements, home range, diet, birth and death rates, and other aspects of a rhino's solitary existence. He noted that they prefer *Saccharum spontaneum* to any other grass and that the seeds of *Trewia* that pass in their dung colonize the floodplains with stands of these trees. Most researchers would then have gone home, perhaps written a doctoral dissertation, and moved on. But Eric had become dedicated to the rhino's future. He realized that offering protection, though absolutely essential, is not enough when thousands of villagers crowd the park's border and demand access to resources and when rhino horn sells for as much as $30,000 a kilogram.

Effective conservation requires partnerships with the local people, as is now generally realized, for they must ultimately be motivated enough to guard and manage their resources. Such motivation comes primarily from economic incentives. To provide these, various organizations have uncritically promoted local control of resources, even inside reserves. As we have learned with dismay, this does not lead to sustainable use; universal will urges individuals to satisfy goals that can never be satisfied. Greed exceeds need.

Eric Dinerstein realized that the local people could not manage resources on their own, that they need advice and technical assistance. They had to find ways to raise money and to derive greater benefits from their environment without intruding on the park. With patience, empathy, and insight Eric and his Nepalese co-workers, as well as a few innovative local people, promoted various conservation ideas. Fenced tree plantations provided timber, fuelwood, and livestock fodder. Public awareness campaigns helped build a local constituency for wildlife. Chitwan is fortunate in that it receives many Nepalese and foreign tourists, but little of that money reached local people until legislation mandated that park fees had to be divided more equitably. A remarkable achievement was that villagers have turned some of their land — 30 km^2 — to forest. Rhinos and tigers recolonized, tourists came to these community forests, and the local people now reap all the benefits. When overfishing ceased, crocodiles returned and once again bask on riverbanks, offering another tourist attraction.

Of course, every reserve, every landscape has its unique problems and requires innovative solutions. However, this elegant case history of Chitwan shows that one individual can have a major influence on conservation if he or she makes a tenacious commitment. Results such as these provide a "kernel of hope," to quote the author. Hope tends to transcend experience. But Chitwan does give hope and a measure of optimism.

We must, all of us — scientists, officials, villagers, urban residents — fight for a noble future; we must begin with a new strategy for survival in order to prevent the splendors of the natural world from dying in our hands. The rhinos of Chitwan are unaware of their precarious existence. Their fate depends wholly on us, on our commitment to protect them forever.

George B. Schaller
Wildlife Conservation Society

PREFACE

AS A CHILD GROWING UP in New Jersey, my field observations of rhinoceros were limited to museum dioramas and a vivid imagination. My real introduction to large Asian mammals came more than a decade later when, in 1975, I was accepted into the Peace Corps and sent off to study the prey species of tigers and their habitats in what is now Royal Bardia National Park in western Nepal. In 1984 I finally had the privilege of living out my childhood fantasies when I was offered a postdoctoral fellowship by the Conservation and Research Center of the National Zoological Park to study rhinoceros in Royal Chitwan National Park, Nepal. The four years of fieldwork that ensued form the core of this book. Since 1989, I have returned annually to Nepal to continue work on large mammal conservation and to maintain my lifelong attachment to the wildlife, natural landscapes, and people of Nepal.

This book is as much about my own transformation as it is about rhinoceros: from a child who dreamed of studying prehistoric large mammals to a curious wildlife biologist who counted rhinoceros and picked through their dung to a conservation biolo-

gist focused on saving an endangered species and its habitat. I tried to follow the advice of George Schaller in writing this book: "Natural history provides the basis for understanding the requirements of a species, something
that researchers often disregard in this age of remote sensing and computer
modeling. Yet describing the life of an animal is not enough: a researcher
today also has a moral responsibility to help the species endure."

I have attempted to distill more than two decades of observations on
Asian wildlife into a conservation strategy applicable to much of the region's dwindling megafauna. Above all, I have tried to highlight a little-
known success story of the recovery of an endangered large mammal population in a human-dominated landscape. I hope this story will be of value
to everyone who has ever gazed in awe at rhinoceros, be it in the wild, in a
zoo, or, as I once did, in a museum diorama.

Many people have inspired me both directly and indirectly to write this
book. George Schaller's landmark study, *The Deer and the Tiger*, was my
bible as a Peace Corps volunteer in Royal Bardia National Park. Claire Dyck-
man inspired me to join the Peace Corps and go to Nepal. Andrew Laurie
willingly shared his insights on rhinoceros biology in Chitwan and showed
me how to track them. Gagan Singh was my first jungle mentor and Nepalese field companion and taught me the names of Bardia's plants and animals. Few have contributed more to the conservation of Royal Bardia National Park and the protection of its recently translocated rhinoceros
population than Gagan. Will Weber, my Peace Corps training director, and
fellow Peace Corps biologists John Lehmkuhl, Luke Golobitsch, Clifford
Rice, Tom Dahmer, and Paul Wilson shared their knowledge of natural history and were great field companions. Narayan Khazi Shrestha, Chij Kumar
Shrestha, and their colleagues taught me the intricacies of Nepali, which has
proved to be a lasting gift. After my Peace Corps service my graduate ad-
viser, Richard D. Taber, welcomed a naive graduate student to the University of Washington and gave me the guidance and freedom to continue work
overseas. Gordon Orians, Vim Wright, P. Dee Boersma, Kai Lee, and other
colleagues at the Institute for Environmental Studies provided support and
served as role models in meshing environmentalism, science, and policy,
while I prepared to return to Nepal to continue fieldwork.

My biggest debt of gratitude goes to Chris Wemmer, director of conservation at the Conservation and Research Center, U.S. National Zoological Park, who invited me to join the Smithsonian/Nepal Terai Ecology Project. Together with Ross Simons and David Challinor of the Smithsonian

Institution, he supervised an eighteen-year wildlife research program that has made Chitwan the best-studied wildlife reserve in Asia. Hemanta Mishra, my co-investigator on the project and the first director of the King Mahendra Trust for Nature Conservation, helped in innumerable ways to make the research project and rhinoceros conservation a success in Nepal. The recognition that he earned as a recipient of the Getty Prize for conservation is a testament to his visionary leadership, which has catapulted Nepal to the forefront of wildlife conservation efforts worldwide. Sushma Mishra educated me about Chitwan and Nepal. The Department of National Parks and Wildlife Conservation gave me permission to live and work in Chitwan, and I would like to particularly thank Directors General Tirtha M. Maskey, Uday R. Sharma, and Biswa N. Upreti and Deputy Director General Bijay Kattel for their support and encouragement. Chitwan's chief warden, Gopal Upaidiya, and former warden, Ram Pritt Yadav, have provided inspiring leadership for landscape conservation activities. Jay Pratap Rana, Chandra Gurung, and Arup Rajuria, as member secretaries of the King Mahendra Trust, supported these efforts.

I collaborated with many Nepalese biologists and field staff for the research and conservation activities that form the basis of this book. In particular, I gratefully acknowledge the assistance of the field staff of the King Mahendra Trust — Vishnu B. Lama, Man B. Lama, Harka M. Lama, Man Singh Lama, Bul Bahadur Lama, Ram Kumar Aryal, and Kesab Giri — in the biological monitoring program. Dr. Sundar Shrestha, chief wildlife veterinarian, supervised the capture and immobilization of rhinoceros. Shant Raj Jnawali and Anup R. Joshi conducted their graduate research on rhinoceros biology under my supervision; their theses, on diet in greater one-horned rhinoceros and seed dispersal of nonfleshy-fruited plants, respectively, figure prominently in this book. The staff members of the trust's elephant stable were willing companions during all aspects of the field research and particularly during twenty-four-hour activity watches of rhinoceros. Their fearless efforts and good humor under difficult conditions were a constant lesson in humility. I would like to mention the following elephant staff members for their support: Badhai Lal Subbha, Gyan Bahadur Tharu, Phirta Tharu, Bir Bahadur Lama, Brij Lal Tharu, Pashupat Tharu, Badri Tharu, Ram Raj Tharu, Arjun Kumal, Maila Kumal, Ram Bahadur Gurung, and Ram Ji Tharu. B. M. Shrestha of the Kumaltar Agriculture Station, Kathmandu, maintained the liquid-nitrogen tank that held blood samples from captured rhinoceros. The scientists and staff at the Rampur Agricultural Station allowed me to use their facilities to process plant samples for nitrogen and detergent fiber analyses. John Lehmkuhl was a valuable coresearcher, and his grassland ecology studies have contributed to this effort. Lori Price helped census rhinoceros and developed the photographic identification library. Dan Janzen inspired me to study rhinoceros–plant interactions and gave sound advice when my spirits sagged. Linda Kentro, James Giambrone, Brian and Judy Hollander, Elliot Higgins, Claire Burkert, Wendy Cronin, Brot Coburn, David Shlim, and Sue Praill offered moral support.

Had this book been completed in 1989, as I originally intended, it probably would have had a more pessimistic message about the future of rhinoceros and the wildlife parks of Asia. The hard work of many of my colleagues now offers the foundation for the hope that poor villagers can share landscapes with large mammals in perpetuity. One role I expect this book to play is to publicize to a larger conservation community the contributions of many dedicated Nepalese conservationists. Much of the success of the habitat regeneration program in the Chitwan buffer zone, particularly in the early days, must be credited to Shankar Choudhury. In 1988 he convinced all skeptics, including Hemanta Mishra and me, that a simple tree nursery program could lead to the regeneration of the buffer zone. His efforts are testament to the belief that one committed person can help restore a landscape. Arup Rajuria, Arun Rijal, Top Bahadur Katri, Ram Kumar Aryal, and Narayan Poudhel have helped direct the efforts to extend this landscape. Kapil Podrel, along with the other field technicians for the trust, monitored the recovery of rhinoceros, tigers, and bird species in the habitat regeneration areas of the buffer zone. Hank Cauley, as the director of the Biodiversity Conservation Network of the U.S. Agency for International Development (USAID), provided equal parts financing and goading to push conservation activities forward in the buffer zone. Marnie Bookbinder Murray, a research scientist, helped design and evaluate the buffer-zone recovery program and the analysis of the effects of ecotourism on local stakeholders. I would like to thank the 30,000 people of the Bagmara, Kumrose, and Kothar User Group Committees for extending protection to 5 tigers, 55 rhinoceros, and the populations of 191 bird species that now use the 30 km^2 of buffer zone under their management.

The staff of the Conservation Science Program of World Wildlife Fund–United States helped in innumerable ways. David Olson, Eric Wikramanayake, George Powell, and Sue Palminteri provided stimulating discussions about the biology and conservation of Asian mammals. John Lamoreux, Patrick Hurley, Emma Underwood, Colby Loucks, Tom Allnutt, and Wes Wettengel helped prepare maps and figures and offered essential advice. Meseret Taye showed extreme dedication and kindness by retyping tables, rearranging figures, and saving me from giving up on a number of occasions. The staff of the Asia-Pacific Program of the World Wildlife Fund, notably Bruce Bunting, Martha van der Voort, and Mingma Norbu Sherpa, provided encouragement and commitment to rhino conservation. Diane Wood and Jim Leape allowed me the solitude to complete the first draft. Ginette Hemley, vice president of the species conservation program of the World Wildlife Fund–United States, and her colleague Steve Osofsky have been strong supporters of efforts to conserve endangered species and of the project in Chitwan. Judy Mills, of the same program, educated me about the myths and realities of the consumption and trade in rhinoceros horn. Kathy Saterson, Burt Levenson, and Daniel Miller of USAID helped in the design and funding of the buffer-zone program and made important contributions to the project.

Drafts of manuscripts that formed the basis of the book chapters benefited from reviews by Jacques Flamand, Daniel Janzen, Henry Howe, Joel Berger, Ted Grand, Andrew Laurie, Katherine Ralls, Chris Wemmer, Don Wilson, John Robinson, Kent Redford, David Challinor, Ross Simons, Jon Ballou, Dave Dance, Leeann Hayek, Mitch Bush, Kathy Saterson, John Lehmkuhl, Richard Taber, Peter Brussard, John Gittleman, Stuart Pimm, Carol Augspurger, Shant Raj Jnawali, Anup Joshi, Per Wegge, Paul Wilson, Clifford Rice, Dave Smith, Charles Janson, Nathaniel Wheelwright, Raman Sukumar, AJ Johnsingh, Eric Wikramanayake, Joy Belsky, Thomas Hanley, Robert Marquis, Gordon Orians, John Thompson, Daniel Edge, Mel Sunquist, Thomas Foose, Gary McCracken, Mark Schaffer, Ted Stevens, Robert Stevenson, and Steve Thompson. Sheila O'Connor, Robin Abell, and Eric Wikramanayake reviewed the entire manuscript. Holly Strand served as a patient, wise editor of several drafts. Andrea Brunholzl also edited the manuscript and was a frequent source of inspiration, along with Seamus and Declan. My sister, Holly Dinerstein, and my mother, Eleanor Dinerstein, patiently reminded me about finishing this book. I owe them a great deal. George Schaller, Kent Redford, and Mac Hunter provided a critical review of the final draft, and I thank them for their effort.

The recovery of Chitwan's rhinoceros population would not have been possible without the enlightened financial support of many donors. The Smithsonian Institution provided eighteen years of core funding that supported the field research on tigers, their prey species, and rhinoceros that became the foundation for the restoration program. The World Wildlife Fund–United States helped finance translocation and antipoaching programs. Restoration efforts in the buffer zone would not have been possible without timely intervention by Hank Cauley of the Biodiversity Conservation Network; Ed Ahnert, Nancy Sherman, and John Seidensticker of the Save the Tiger Fund; and private donations from Jeffrey Berenson. A grant from the Armand K. Erpf Foundation and from Jeffrey Berenson allowed me to find time away from my duties as chief scientist at World Wildlife Fund–United States to complete this book. My own institution and its president, Kathryn Fuller, supported my efforts to chronicle this story on behalf of World Wildlife Fund's mission to conserve endangered species around the world.

Finally, I acknowledge everyone who has ever studied large Asian mammals and whose scientific papers and books have educated me about their biology. Those of us who have lived among and studied large Asian mammals are a privileged few. But with that privilege comes a responsibility to ensure that some fraction of the earth remains to conserve these magnificent animals.

the return of the unicorns

INTRODUCTION

ON A MISTY FEBRUARY MORNING in 1986, I sat patiently in the fork of a kadam tree in Royal Bardia National Park in southwestern Nepal. The forest echoed with the raucous calls of peacocks and wild jungle fowl, but my attention was fixed on two enormous crates at the edge of the grassland. Inside the crates were two greater one-horned rhinoceros, a pair captured a day earlier in Royal Chitwan National Park, Nepal, 250 km to the east. I was waiting to photograph their release, marking the end of two years of intensive research and planning that had culminated in the return of rhinos to sanctuaries within their former historical range. After a few moments of hesitation, the rhinos rushed out of their crates to reclaim their lost territory. Rhinoceros probably disappeared from Bardia 100 years ago. Today the park is well protected and contains enough habitat to support a sizable population.

Between 1986 and 1991, the translocation program, a combined effort of the government of Nepal, Smithsonian Institution, and World Wildlife Fund, brought about the successful transfer of thirty-nine animals. By early 1991, five females had given birth to calves that survived their first year. The program continued, and by April 2001 many more rhinoceros had arrived in Bardia, bringing the total of translocations to seventy-eight. And for the first time, the program sent four individuals to the Royal Sukla Phanta Wildlife Reserve in far western Nepal. This marked the second translocation to a prime location within the species's former range, and the beginning of another chapter in one of Asia's biggest conservation success stories.

The drive back to Chitwan would always temper my personal elation during these periodic translocations. We drove along the national highway in lowland Nepal, through poverty-stricken villages huddled along the roadsides, dry rice paddies, and degraded pastures and forestlands over-stocked with goats and cattle. In the midst of this dusty landscape of rural poverty, a nagging question lingered: What future do large mammals and wildlands have in densely populated Asia?

High rates of both land conversion and population growth threaten conservation of wildlife habitats. In Asia, 50% of the world's population crowds into 13% of the land area. In response to the rapid loss of wildlife habitats, the governments of Asian nations have established more than 1,000 reserves to conserve biological diversity. Efforts to conserve large mammal populations have strongly influenced the selection of Asian reserves for pro-tection. However, a recent study of reserve size in this region shows that Asian parks are considerably smaller than reserves in Latin America or Africa and that many may be too small to conserve viable populations of large mammals over the long term (Dinerstein and Wikramanayake 1993).

The bleak prospects for conservation of large Asian mammals are well documented. Book titles such as *Stones of Silence* (Schaller 1988) and *The Twilight of India's Wildlife* (Seshadri 1969) are suggestive of the precarious status of many large mammal species. The three species of Asian rhinoceros are limited to relatively small, isolated protected areas because much of their original habitat has been destroyed. Furthermore, poachers have wiped out many populations. The most endangered of the three species, the Javan rhi-noceros, is one of the rarest mammals on Earth, with only two populations that together represent no more than eighty individuals (Foose and van Strien 1997). A recent review of tiger populations identified 159 areas, or tiger conservation units, in the Indian subcontinent, Indochina, and Suma-tra where tigers are known or thought to occur (Dinerstein et al. 1997; Wikramanayake et al. 1998a, 1999). Perhaps no more than 30 of these units will continue to maintain healthy tiger populations through 2020. While much of the world's attention is fixated on African elephant populations, estimated at 650,000 animals, Asiatic elephant populations are in grave danger. Perhaps no more than 35,000 Asiatic elephants remain in the wild; worse, only ten populations may number more than 1,000 individuals (Sukumar 1999). Populations of large mammals in lowland areas are particularly vulnerable because these habitats are under the greatest pressures from conversion to agriculture and oil-palm plantations (Wikramanayake et al. 2002). Fragmentation and isolation of reserves increase demographic threats of extinction for some species.

Can we be optimistic about conservation efforts for large mammal populations in the human-dominated landscapes of Asia? This book provides a kernel of hope in the struggle to save the vanishing large mammal species of the region. My central purpose is to extol a poorly publicized success story in wildlife conservation: the recovery of an endangered large herbivore and its habitat in one of the world's poorest countries. I will show how a recovery program built on strict protection, targeted field research, landscape-scale planning that includes economic incentives to promote local guardianship of endangered wildlife and habitats, and bold leadership can lead to the resurgence of even the most endangered large mammals.

After my introduction to large Asian mammals as a Peace Corps volunteer at Royal Bardia National Park in 1975, I conducted fieldwork on rhinoceros and other large Asian mammals in three phases and in different capacities. As a biologist working under the guidance of the Conservation and Research Center of the U.S. National Zoological Park, I completed a detailed biological study, mostly within Royal Chitwan National Park, from 1984 to 1988. As a scientist working for the World Wildlife Fund, I augmented my experiences in Bardia and Chitwan with visits, taught field courses, and collaborated with local scientists in a number of other large mammal reserves across Asia from 1988 until 1994. Since 1994, my fieldwork has focused on the recovery of endangered species and habitats in the buffer zone and larger landscape surrounding Royal Chitwan National Park. As chief scientist of the World Wildlife Fund–United States, my most recent concern has become the design of the Terai Arc, a regional network of linked protected areas, to manage rhinoceros, tigers, and elephants as metapopulations (populations linked by dispersal); I discuss this topic in chapter 11.

The initial goal of my 1984 field project was to study rhinoceros–plant interactions. This study was inspired by Dan Janzen's work in the tropical dry forests of Costa Rica, where he used large domesticated ungulates as surrogates for native megaherbivores. I soon learned that the density of the vegetation often made locating the animals impossible. Fortunately, the Department of National Parks and Wildlife allowed me to capture, immobilize, radio-collar, and measure rhinoceros on a regular basis. This opened new opportunities for research in rhinoceros–plant interactions and virtually all aspects of their biology. Protocols developed during capture operations (Dinerstein, Shrestha, and Mishra 1988) made the ambitious translocation program possible, and it subsequently became an important part of our work. Earlier studies of Asian rhinoceros did not have the advantages of radiotelemetry because government officials were unwilling to permit the immobilization of endangered species, some populations were rare, and in the 1970s and 1980s conducting telemetry studies on wide-ranging vertebrates was both costly and fraught with logistical difficulties. We gained a huge advantage because we could habituate radio-collared animals to close approach by observers on elephant-back.

The needs of local people and the effects of human settlements on endangered habitats are a fundamental part of landscape-scale conservation. By 1994, my fieldwork was focusing as much on human behavior, ecotourism, and local management of natural resources as on observing rhinoceros. How the field program evolved into these new areas is the basis for part III.

The sequence of the book chapters emphasizes a conservation biology approach: describing the status and root causes of decline of an endangered species (part I), understanding basic biology (part II), and applying biological insights and knowledge of socioeconomic conditions to the design of conservation programs (part III). Each chapter begins with an anecdote about my experiences while studying rhinoceros or other large Asian mammals. Each chapter ends with statistical notes, where applicable. Appendix A describes the methods used in the field study.

Part I, " Vanishing Mammals, Vanishing Landscapes," covers the demise of rhinoceros and their habitats and identifies the causes of the loss. Chapter 1 traces the evolution and decline of rhinoceros and their relatives and looks at why they have persisted for so long. This chapter also covers the current status of remaining populations of rhinoceros in Asia. Chapter 2 examines the threats to rhinoceros caused by demands for its body parts, notably its fabled horn. Chapter 3 introduces the endangered floodplain ecosystems of the study area and provides a brief history of conservation of Nepalese megafauna.

Part II, "The Biology of an Endangered Herbivore," explores the rhino's natural history, providing the scientific foundation necessary for devising conservation plans. Within this section, I often refer to Royal Chitwan National Park; with its rich megafauna, Chitwan is a living laboratory—a window on the Pleistocene—where one can observe and study interactions between mammals and their environment not unlike those that occurred when giant herbivores dominated the landscape. Chapter 4 begins with a description of the physical characteristics of greater one-horned rhinos and a comparison of them with other rhinoceros in terms of both size and degree of sexual dimorphism. I show how extremely large body size influences various aspects of rhinoceros ecology and impinges on conservation efforts. Chapter 5 challenges certain aspects of the argument that very large herbivores are by nature prone to extinction. I begin by describing the demographic, genetic, and environmental threats faced by rhinoceros. Then I trace the rapid recovery of the Chitwan population from a population bottleneck, and I discuss the remarkably high levels of genetic variability in the Chitwan rhinoceros population and the implications of these findings for other species. Chapter 6 focuses on the effects of very large body size on many aspects of rhinoceros biology, including use of space, feeding ecology, activity patterns, and thermoregulatory strategies. The social organization of an endangered species is extremely important because it ultimately affects such issues as population management and reserve design. Chapter 7 relates biology to social behavior by describing the rapid turnover of dominant males in areas of high densities of breeding females; it also discusses how the size and condition of the tusks (i.e., procumbent outer incisors), rather than the horn or other secondary sexual characteristics, largely determine dominance. In chapter 8, I illustrate the role that giant herbivores play as landscape architects while highlighting just how the rhinos do this. In some parts of their range, giant herbivore populations have been so decimated that they have suffered ecological extinction; they may persist in small numbers, but their once-prominent role in the ecosystem has vanished. In contrast, the presence of large prehistoric herbivores in Chitwan allows us to evaluate evolutionary theories about how giant mammals and plants interact without having to experiment instead with surrogate species such as domesticated livestock.

Field studies detailing the fascinating story of rhinoceros as megafaunal seed dispersers may interest evolutionary biologists but mean little to subsistence-level farmers, who view rhinos as unwanted consumers of their crops. Part III, "The Recovery of Endangered Large Mammal Populations and Their Habitats in Asia," devotes considerable attention to projects that attempt to link large mammal conservation and, by extension, biodiversity protection, with local development. I begin, in chapter 9, by focusing on the role of ecotourism, using large mammals as the target species, and I document the effects of twenty years of ecotourism on conservation in Chitwan. Chapter 10 evaluates a highly controversial subject in conservation biology — the utility of integrated conservation and development projects, also known as eco-development projects or sustainable use projects. These projects attempt to simultaneously raise local living standards while conserving endangered species and their habitats. I explore the reasons why these projects typically fall short of their objectives. One conclusion is that, absent strict protection of large core areas, eco-development projects are bound to fail. Chitwan is a successful example of eco-development. This program has allowed local people living outside the park to become local guardians of endangered species and to participate more actively in habitat regeneration programs.

My final objective is to promote an integrated conservation strategy for Asian megafauna. In chapter 11, I propose that, given adequate protection from poaching and provided with suitable habitat, even some of the largest, slowest-breeding mammals can recover quickly from episodes of near extinction. The final chapter introduces an experiment in landscape-scale conservation for rhinoceros, tigers, and wild elephants — the Terai Arc — that draws on the natural history studies that I describe in part II and builds on the lessons learned from the conservation program. The Terai Arc is designed to link eleven national parks and reserves across southern Nepal and northern India through wildlife corridors. The Terai Arc, which was being implemented as this book went to press, will, I predict, be viewed as the most ambitious wildlife recovery project in Asia.

I hope that readers will end this book with a thorough appreciation not only of the biology of this fascinating mammal but also of the endangered habitats for which rhinoceros serve as a flagship species. The vanishing landscape of lush floodplain grasslands and riverine forests at the base of the Himalayas is one of the most productive areas for wild ungulates on Earth and one of the most threatened. Ultimately, the conviction to conserve endangered large mammals must spring from an appreciation of their intrinsic value and widespread public awareness, especially of the reality that once extinct, species are gone forever. I hope that, in its own small way, this book spurs concerned individuals to speak up for the large Asian mammals that have no voice in their future.

PART I

Vanishing Mammals, Vanishing Landscapes

TWENTY-FIVE MILLION YEARS AGO, on a warm, calm day in the Oligocene, a male giraffe-rhinoceros (*Paraceratherium grangeri*) ambled through a shaded grove at the base of a low range of hills in South Asia, clipping the leaves of tree branches with his massive teeth. Twenty feet long and as tall as a double-decker bus, the giraffe-rhinoceros was the largest land mammal to tread the planet, dwarfing even prehistoric elephants and mammoths. Nearby, a female giraffe-rhinoceros and her two-year-old calf nibbled on some low-hanging foliage. The mother stopped to listen, alert for giant hyena-like predators that might attack her offspring. Soon the calf would outgrow the danger of being taken by the dog- and catlike creatures that, along with the hyenas, were the dominant predators of their day. Meanwhile, primitive giant pigs rooted along the forest's edge, and small groups of the earliest ancestors of deer and wild cattle grazed in the sunlight. This was truly a world where giant mammals dominated the earth.

Today, at the base of the Siwalik Range in lowland Nepal and India, the giraffe-rhinoceros and other giant mammals of South Asia can still be found, but not in the national parks or wildlife reserves. They appear only as fossils in the treasure chest of old mammal bones that forms part of the Siwalik formations. The vast floodplains at the base of these mountains are still there, but they support rice culture and exploding human populations, with only a few swaths left for the remaining megafauna, the giant mammals exceeding 300 kg in body mass, that are still with us. Rhinoceros, elephants, deer, forest cattle, pigs, and the large predators that hunted them once flourished here but now are gone forever. What remains is a handful of a once-vibrant evolutionary stable of large mammals.

Some scientists hold out hope for re-creating this megafauna through genetic engineering, re-creating a kind of "Oligocene Park" or "Pliocene Park." But if their test-tube woolly rhinoceros or mammoths fail to materialize, we are left with the painful truth: that for the rest of the history of the planet, the only giant terrestrial herbivores that will still be around are two species of elephants, five species of rhinoceros, hippos, eight species of bear, several species of wild cattle, and giraffes. Of course, even if ambitious reproductive biologists were successful at breeding and cloning big mammals, viable populations of most of these prehistoric species would have a hard time squeezing into the small reserves allocated for nature protection. In essence, we have become caretakers for elderly species in a high-rent landscape.

CHAPTER 1

VANISHING MAMMALS:
THE RISE AND FALL OF THE RHINOCEROS

We must face the fact that the Cenozoic, the Age of Mammals, which has been in retreat since the late Pleistocene, is over, and that the "Anthropozoic" or "Catastrophozoic" has begun. Our task now is to salvage some samples of the megafauna and protect enough habitat to give future human beings an opportunity to restore a semblance of evolutionary integrity in the 22nd century.
—Michael E. Soulé, "The End of Evolution?"

MY FIRST VIEW OF A GREATER one-horned rhinoceros was at the age of seven in the Hall of Asian Mammals at the American Museum of Natural History in New York. My classmates and I, a group of boisterous fourth-graders from the Toms River, New Jersey, Elementary School, had just finished writing and illustrating a book on the ecology of dinosaurs. We planned to seek critical review of our manuscript from the museum's own Dr. Roy Chapman Andrews, the dean of American paleontologists, and our hero. After wandering through the dinosaur exhibits all morning, I broke off from the group to take in some more modern creatures. The diorama of the rhinoceros took me by surprise. I was tall for my age, but I was astounded by the enormous size of the stuffed rhinoceros and the larger-than-life backdrop of its habitat — elephant grasses taller than elephants, brightly colored birds and insects, a new and enticing world fashioned around this rough beast. I felt my allegiances shift almost instantaneously: I wanted to learn more about this animal that seemed as fantastic as any dinosaur yet is still alive.

Of course, the unusually interesting evolutionary history of the rhinoceros rewarded my curiosity several times over. An ungulate, or hoofed mammal, the rhinoceros belongs to the mammalian order Perissodactyla (from Greek *perissos*, of numbers odd; *daktulos*, a finger or toe), together with horses and tapirs. Of the five living species, two occur in Africa—the black rhinoceros (*Diceros bicornis*) and the white, or square-lipped, rhinoceros (*Ceratotherium simum*)—and three species in Asia: the greater one-horned rhinoceros (*Rhinoceros unicornis*), the Javan rhinoceros (*Rhinoceros sondaicus*), and the Sumatran rhinoceros (*Dicerorhinus sumatrensis*).

Most of the recent popular and scientific literature on rhinoceros has focused on their rapid demise. All five species have undergone drastic reductions in distribution and numbers since the mid-nineteenth century.

The Evolution of the Rhinoceros

Three features mark the evolutionary history of rhinoceros: the antiquity of the lineage, the diversity and variety of the ungulate feeding niches that they occupy, and numerical abundance.* The earliest known rhinoceros-like mammal belongs to the genus *Hyrachyus*, known from late Eocene deposits in Asia, North America, and Europe (Prothero, Guerin, and Manning 1986). These primitive rhinoceros resembled early horses and tapirs—diminutive, delicate, with no horns. Not until later in their evolutionary history did horns become a defining feature of rhinoceros. The first true rhinoceros that resembled modern forms belong to the family Rhinoceratidae. They also appeared in the late Eocene but were less dominant than two other families: the running rhinoceros (Hyracodontidae) and the aquatic rhinoceros (Amynodontidae) (figure 1.1).

The Hyracodontidae are divided into two groups: a dog-size cursorial

---

*Paleontologists have inadequately reconstructed the evolutionary lineage of rhinoceros; mistakes in nomenclature, and misconceptions about rhinoceros systematics and phylogenies, are common in the scientific and popular literature until 1988 (Prothero, Guerin, and Manning 1989). Older phylogenies were based on features now known to be highly variable (molarization of the premolars), resulting in the splitting of primitive rhinoceros into too many species. The synopsis of rhinoceros evolution in this chapter is drawn from the most recent reviews (Prothero, Guerin, and Manning 1989; Prothero and Schoch 1989).

FIGURE 1.1 Various forms of ancient perissodactyls. The ancestral form of Perissodactyl was quite small and lacked horns (*Hyrachyus, lower right*). The giraffe-rhinoceros (*Paraceratherium*) was, in its day, the world's largest land mammal. Both the Etruscan rhinoceros (*Dicerorhinus etruscus*) and the woolly rhinoceros (*Coelodonta*) are related to the living Sumatran rhinoceros. (Reprinted with permission from Wildlife Education, Ltd., *Zoobooks*)

version (the Hyracodontidae) and an immense form, or lineage (the Indricotheriinae). The latter group included some truly spectacular rhinoceros. For example, the giraffe-rhinoceros (*Paraceratherium grangeri,* formerly known as *Baluchitherium,* or *Indricotherium grangeri*) of Mongolia (Lucas and Sobus 1989) was the largest terrestrial mammal. It was an impressive sight, with a shoulder height of almost 6 m and approaching a total length of 9 m (Osborn 1923). The indricotheres, including the giraffe-rhinoceros, disappeared from Asia by the middle Miocene.

Rhinoceros are considered biologically successful because of both their evolutionary persistence and their widespread distribution. The aquatic rhinoceros reached their maximum diversity in the late Eocene and early Oligocene, especially in Asia (Wall 1989). The evolutionary persistence of the representatives of this family was remarkable. For example, the hippo-like rhinoceros (*Metamynodon*) ranged for 10 million years over much of North America. By the early Oligocene, most aquatic rhinoceros were in decline, with one genus, *Cadurcotherium,* surviving until the middle Miocene in what is now Pakistan, nearly 15 million years after the demise of the other aquatic rhinoceros.

The third family of rhinoceros, the Rhinocerotidae (figure 1.2), flourished in the Oligocene, after first appearing in the late Eocene in Eurasia. Small species came first, followed by middle- and large-size rhinoceros during the Upper Oligocene. Interestingly, Africa lacked rhinocerotids during the same period. In contrast, North American rhinoceros increased in size and diversity much earlier than in Eurasia and featured the first rhinoceros with horns, *Diceratherium*. *Diceratherium* also exhibited persistence and dominance. For nearly 10 million years, it was the only megaherbivore in North America.

These lineages exhibit several evolutionary trends observed in mammals. Perhaps the most dramatic is the increase in body size, as formalized by Cope's rule (1880), which theorizes that species within a lineage tend toward gigantism over evolutionary time (Heissig 1989). However, incidences of dwarfism, particularly in the hippolike genus *Teleoceras,* also occurred. Another development in the lineage was the appearance of boneless horns on the skull, but horn size, number, and placement have varied consider-

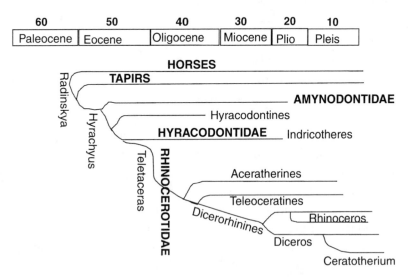

FIGURE 1.2 Family tree of the major groups of rhinocerotoids and related perissodactyls. This family tree illustrates presumed phylogenetic relationships, divergence, and extinction. The geologic time scale is in millions of years; Plio, Pliocene; Pleis, Pleistocene, including the Ice Age. Today only five species of rhinoceros remain in one of the most diverse groups of herbivorous mammals to appear on Earth. (Adapted from Prothero 1993:91)

ably. The evolution of broad feet with three toes became another trade-
mark. Less conspicuous, but of considerable ecological importance, were
changes in the dentition: the premolars became more like molars, the
crowns of the cheek teeth lengthened, and the enamel patterns became
more complex. These changes increased efficiency in handling a diet of
coarse grasses for those species that were primarily grazers.

Climatic changes probably triggered the demise of many groups of rhinoceros at the end of the Miocene. In North America, almost the entire rhinoceros fauna was eliminated, and in Eurasia only two lineages survived.
One of these, the dicerorhinines, gave rise to a form very similar to the
Sumatran rhinoceros of today. Perhaps the most famous member of this
lineage was the woolly rhinoceros (*Coelodonta antiquitatis*), which appeared in the Pleistocene in China and moved westward into Europe. In the
Upper Pleistocene, the woolly rhinoceros ranged from Korea to Spain, the
widest range of all species recorded.

Today, rhinos inhabit only Africa and Asia. Africa's black rhinoceros
first emerged 4 million years ago in the Pliocene. Until the last few hundred
years, this species ranged throughout sub-Saharan Africa. These features establish the black rhinoceros as one of the oldest, most stable, and widely distributed species in the contemporary African megafauna. The other living
African species — the white rhinoceros — appeared more recently, in the
middle Pleistocene. Of the living Asian rhinoceros, the Sumatran rhinoceros has changed little from its Miocene ancestors, which lived more than
40 million years ago. The Sumatran rhinoceros is related to the woolly rhinoceros and may be more closely related to the African species than to the
two Asian species, which have a single horn. The greater one-horned, or Indian, rhinoceros and the Javan rhinoceros appeared in the middle Pleistocene.

Occupation of virtually all ungulate feeding niches reflects the diversity
of the rhinoceros. Some of the earliest ones were small grazers, and others
were small browsers. The giraffe-rhinoceros browsed the treetops, and an
assortment of other species grazed or browsed shrubs and saplings in the
understory. The semiaquatic, hippolike *Teleoceras* is one of a number of
species that fed on aquatic plants common to rivers, streams, and oxbows.
Other rhinoceros plucked vegetation with their tapirlike snouts. The extant
greater one-horned rhinoceros, although adapted to a diet of grasses and
browse, eats large amounts of fruit seasonally — a phenomenon that probably occurred in other forest-dwelling species. Perhaps the only niche be-

haviors that the rhinoceros did not exploit were digging and rooting or foraging on bottom plants in lakes.

Their ability to exploit a variety of habitats explains in part the cosmo-
politan distribution of rhinoceros. Rhinoceros could be found both on flat
terrain and in hilly rugged country. Species shared the cold steppes with
mammoths, the hot dry savannas of Africa with antelopes, and the humid
Asian forests with elephants. From the Oligocene to the Pleistocene, nearly
all terrestrial habitats in the Northern Hemisphere contained one or more
rhinoceros species as part of their mammal fauna. In the Northern Hemisphere and in Africa, rhinoceros once occurred in great numbers. In some
habitats, they may have outnumbered some of the smaller ungulates. This
aspect is remarkable because rhinoceros are among the largest herbivores
among Tertiary period vertebrate communities.

However one chooses to define the evolutionary prominence of a lineage — persistence, geographic range, diversity of feeding niches, variety of
habitats used, numerical abundance — the rhinoceros stand as one of the
most successful groups of mammals. Large body size and high mobility
emerge as the most prominent factors in their success. High mobility provided a natural escape against changing climates and forage conditions and
permitted movement to new areas and habitats. It also helped rhinoceros
cope with local, periodic, and often severe disturbances caused by floods
and fires. Among rhinos today, the greater one-horned rhinoceros is a
strong swimmer. Interestingly, the white rhinoceros is far more reluctant to
cross water barriers (Owen-Smith 1973) and occasionally drowns (Jacques
Flamand, personal communication, 1999). Javan rhinoceros are most likely
strong swimmers, but we know little about the ability of Sumatran rhinoceros to cross open water. Both African species and Sumatran rhinoceros are
undaunted by hilly terrain and thus able to cross physical barriers that restrict smaller species.

Large size also offered an effective defense (at least as adults) from predators. Daggerlike teeth (mandibular outer incisors), found in the Asian rhinoceros and probably present in some of the earlier rhinocerotids, provided
an even more effective antipredator defense when combined with large size.
Among living rhinoceros, greater one-horned rhinoceros are subject to predation by tigers (*Panthera tigris*) up until the first year but are safe after that.
However, size alone can be overrated as an antipredator strategy: tigers routinely kill adult male Indian bison, or gaur (*Bos gaurus*) (Ullas Karanth, per-

sonal communication, 1999), which exceed subadult male rhinoceros in size.

The ability to process and subsist on coarse high-fiber plant material must have conferred on rhinoceros a great advantage over smaller ungulates. Most plant communities contain a large quantity of indigestible material or slowly digestible fiber that is relatively low in nutritional content. The digestive strategies of rhinoceros allowed them to subsist on plant matter that would yield too little energy to meet the metabolic demands of smaller herbivores. Because this type of vegetation was so plentiful, the radiation and abundance of large-bodied species able to convert high-fiber diets is not surprising. The development of molars adept at handling coarse grasses, which are often protected by guard cells of enamel-grinding silica, also allowed these large mammals to process the most abundant forage plants.

The ability of some rhinoceros to switch seasonally between browse and grasses may have contributed to their wide distribution. For example, greater one-horned rhinoceros are primarily grazers, and have the high-crowned molars characteristic of this feeding niche, but they can subsist seasonally on a diet of dicotyledonous plants (chapter 6). The ability to shift the proportion of graminoid species and browse probably aided rhinoceros during dispersal into new habitats. Their swimming ability also afforded an opportunity to exploit aquatic plants and extended their flexibility in finding food, as is the case of greater one-horned rhinoceros in Chitwan, which feed extensively on aquatic vegetation during the hot-dry season.

Another reason for their persistence and abundance is their ability to thrive in areas of high habitat disturbance. Today all three surviving Asian species reach their highest densities in early successional habitats maintained by local disturbance regimes. The greater one-horned rhinoceros reaches extraordinary localized densities because it is well adapted to feeding in one of the most dynamic landscapes on Earth — the floodplains of the major river systems of the Indian subcontinent where annual monsoon floods are the norm. Early naturalists described the Sumatran and Javan rhinoceros as pests in the gardens and tea estates of the early colonials in Indonesia. Gardens were part of a highly simplified, disturbed landscape that these large ungulates found attractive. Wild Sumatran rhinoceros seek out forest gaps caused by falling trees, the most common type of disturbance in natural rain forest habitats. Captive Sumatran rhinoceros, for example, prefer food plants such as *Macaranga* (a pioneer tree species in the Euphorbiaceae) that rapidly colonize tree-fall gaps in Asian rain forests.

A final consideration that may have contributed to the long persistence of rhinoceros is its rugged nature. Intraspecific combat is intense among male greater one-horned rhinoceros, and dominant males often inflict serious, sometimes fatal, wounds on subordinate males, estrous females, and subadult males. Research and park staff have also had to treat individuals wounded by other males or speared by angry farmers protecting their crops. Both Laurie (1982) and I have documented instances where we assumed that many of these injured animals would rapidly succumb to their serious wounds, only to watch them return to health in a short time with no apparent sign of infection. There may have been selection for a strong immune response to prevent infection from muscular and subcutaneous wounds. Apparently, male hippopotami, which also engage in violent fights and suffer serious wounds, show a similar propensity to resist infections and recover quickly (Flamand, personal communication, 1999). We know little about intraspecific combat among extinct rhinoceros. Although I have no hard data to support my claim, I suspect that ancestral rhinoceros may have shared the ability of greater one-horned rhinoceros to recover quickly from a majority of serious injuries without signs of infection. If so, the strong immune response may have increased their ability to survive disease outbreaks that might have threatened less robust species.

The Recent Decline and Status of Living Rhinoceros

The contrast between the former dominance of rhinoceros and their current rarity is truly staggering. The current estimate of the total free-ranging population of the three Asian species is fewer than 2,700 individuals (Foose and van Strien 1997). Rhinoceros in Africa have not fared much better, with the exception of white rhinoceros in South Africa, which now holds an estimated 7,100 individuals (Brooks 1994), or 56% of all free-ranging rhinoceros alive today. Even more sobering is the statistic that only two populations of Asian rhinoceros currently contain more than 100 individuals. The black rhinoceros was the most numerous species early in the twentieth century. In Kenya, a current stronghold, black rhinos number only 440 in all, fragmented into fifteen smaller populations, all smaller than 70 animals (Brett 1998). The decline and fragmentation of rhinoceros populations worldwide is alarming because most of the species were abundant until fairly recently and, with the exception of greater one-horned rhinoceros, were widely distributed.

African Rhinoceros

Around 1900, the savannas and woodlands of sub-Saharan Africa may have supported more than 1 million black rhinoceros (figure 1.3). This species may have undergone the most precipitous decline of all living rhinos. Today it is unlikely that more than 2,500 exist in the wild or on game ranches. The four recognized subspecies and countries where they persist are *Diceros bicornis bicornis* in Namibia and South Africa; *Diceros b. longipes* in Cameroon; *Diceros b. michaeli* in Ethiopia, Kenya, South Africa, and Tanzania; and *Diceros b. minor* in Malawi, South Africa, Swaziland, Tanzania, Zimbabwe, and perhaps a few other countries of southern Africa. The largest free-living populations are in Zimbabwe, Namibia, Kenya, and South Africa.

During the colonial era, trophy hunters prized African rhinos. In modern times, hunting is largely prohibited, but poaching pressures remain in-

FIGURE 1.3 An adult black rhinoceros. Until recently, black rhinoceros were the most abundant of all living rhinoceros. (Courtesy Rick Weyerhaeuser/WWF)

tense. To deter poaching, the wildlife officials in Namibia and Zimbabwe routinely dehorned their rhinoceros populations during capture and translocation operations. Kenya has moved some populations to small, well-protected sanctuaries and game ranches, where they have rapidly reached local carrying capacity. One of the most promising efforts to restore the species is in the Selous Game Reserve of southern Tanzania. One of the largest populations on the continent thrived in the territory of this reserve. Now Tanzania is establishing intensive protection zones of roughly 300 to 500 km^2 within the 55,000-km^2 hunting reserve, one of the largest blocks of habitat remaining in East African miombo woodlands.

The white rhinoceros is distributed between two discrete populations. The northern white rhinoceros (*Ceratotherium simum cottoni*) numbered as many as 2,000 animals in 1960, spread over a large area west of the Nile River in northern Uganda, Sudan, the Central African Republic, and the Republic of the Congo. Today poaching has reduced the northern white rhinoceros to a relict population (probably fewer than 30 individuals) confined to Garamba National Park in the Democratic Republic of the Congo (formerly Zaire). Because of the overthrow of Zaire's government in 1997, its future is grim.

The southern white rhinoceros (*Ceratotherium simum simum*) occurs more than 2,000 km to the south (figure 1.4). The northern border of its range was the Zambesi River, and the subspecies extended across to the border of Namibia and Angola. The southern subspecies was virtually wiped out by the end of the nineteenth century; fewer than 100 survived in what is now Hluhluwe-Umfolozi Park, and a smaller population lived in Mozambique. It has rebounded in response to effective protection of more than 7,500 individuals and constitutes a significant African conservation success story. Translocations from Umfolozi to parks in other countries within the historical range suggest a brighter future for this subspecies.

Asian Rhinoceros

The Javan rhinoceros was widely distributed from northeastern India across Indochina to the island of Java in Indonesia (figures 1.5 and 1.6). Estimates of densities are not available, but that Dutch tea planters considered the Javan rhinoceros a pest indicates that they were common in certain parts of their range. Javan rhinoceros appear to prefer lowland forest, but this may be an artifact of their current distribution because their original range in-

FIGURE 1.4 An adult white rhinoceros in South Africa. White rhinoceros are the largest of the five species. (Courtesy Howard Buffett)

FIGURE 1.5 A Javan rhinoceros photographed by infrared camera trap in Vietnam. (Courtesy Mike Baltzer/WWF/CTNPCP)

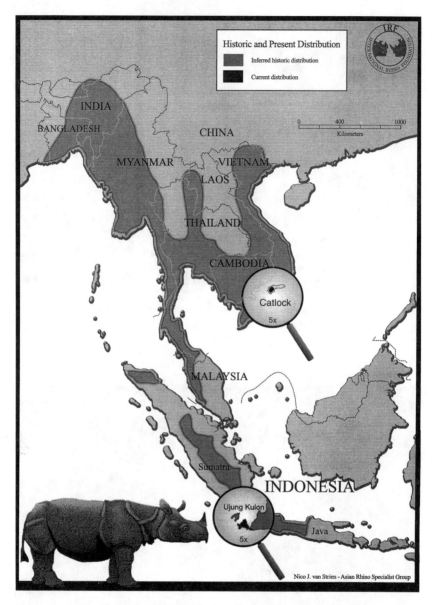

FIGURE 1.6 Historical and current distribution of Javan rhinoceros. (Nico van Strien, International Rhino Foundation)

cludes a number of low mountainous areas. Today Javan rhinoceros are perhaps the rarest species of large mammal in the world and certainly the most
endangered of all five rhino species. The Javan rhino occurs in only two
places: in Ujung Kulon National Park on the westernmost tip of Java, and in
Vietnam. This species is difficult to survey in the wild. Current estimates
based on track counts put the population at fewer than sixty animals in
Ujung Kulon (Foose and van Strien 1997). While conservationists have
worried about the fate of the Indonesian population, the decline in Indochina has been steeper. An area near the Nam Cat Tien Reserve (Cat Loc)
in the southern part of Vietnam near the Vietnamese–Cambodian border
now harbors five to ten animals (Foose and van Strien 1997; David Hulse,
personal communication, 1999) and is the target of concentrated protection efforts.

The Sumatran rhinoceros ranged from Assam, India, and southern
Bhutan south and east to the Indonesian islands of Sumatra and Borneo
(figures 1.7 and 1.8). Today the total Sumatran rhinoceros population numbers fewer than 400 individuals. This figure is based largely on educated
guesses and a few intensive surveys. The last strongholds of large populations all appear to be in Sumatra in Gunung Leuser, Way Kambas, Kerinci

FIGURE 1.7   Sumatran rhinoceros are the smallest and most ancient of the living rhinoceros.
(Courtesy Bruce Bunting/WWF)

FIGURE 1.8 Historical and current distribution of Sumatran rhinoceros. (Nico van Strien, International Rhino Foundation)

Seblat, and Barisan Selatan National Parks. In peninsular Malaysia, the largest population is in Taman Negara, the largest national park in mainland Malaysia. Although Sumatran rhinoceros may have already disappeared from Kalimantan in Borneo, Meijaard (1996) has documented indications of continued inhabitation here. The largest population on Borneo seems to be in the Malaysian state of Sabah in the Tabin Wildlife Reserve, but recent estimates indicate that only a handful of animals persists. Small populations of Sumatran rhinoceros are believed to live at additional sites in Indonesia and Malaysia, but nowhere are the demographic trends encouraging (Foose and van Strien 1997). Sumatran rhinoceros have probably been extirpated from Thailand and Myanmar.

Unlike the other two Asian species, greater one-horned rhinoceros historically were limited to the floodplains and forest tracts in the Brahmaputra, Ganges, and Indus River valleys (figure 1.9). Artifacts from the Mohenjo-Daro era depict greater one-horned rhinoceros, suggesting that the species occupied what is now Sind province in Pakistan in 2000 B.C. (Laurie 1982). These rhinos maintained their extensive distribution until rela-

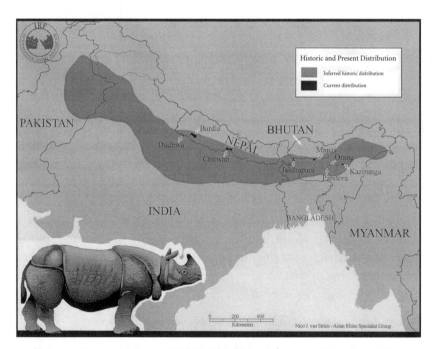

FIGURE 1.9 Historical and current distribution of greater one-horned rhinoceros. (Nico van Strien, International Rhino Foundation)

TABLE 1.1.   DISTRIBUTION AND ABUNDANCE OF EXTANT FREE-RANGING
GREATER ONE-HORNED RHINOCEROS POPULATIONS

Location	Estimated number of rhinoceros 1993	1995	Protection status	Potential carrying capacity
India				
Manas	60+	4	National park World Heritage Site	>100
Dudhwa	11	13	National park	>100
Katarniaghat	4	4	Wildlife sanctuary	10
Kaziranga	1,164 ± 134	1,200	National park	1,500
Laokhowa	0	0	Wildlife sanctuary	50+
Orang	90+	90	Wildlife sanctuary	>150
Pabitora	56	68	Wildlife sanctuary	70+
Pockets-Assam	25	20	Insecure	100+
Jaldhapara	33+	35	Wildlife sanctuary	150+
Gorumara	13	18	National park	50+
Nepal				
Royal Bardia	40+	40	National park	300+
Royal Chitwan	375–400	460	National park	500
Pakistan				
Lal Sohanra	2	2	National park	?
Total	1,870–1,895 ±134	1,948		2,600+

NOTE: For a map of locations, see figure 3.2.

tively recently. Today, however, no more than 2,000 individuals remain in the wild, with only two populations containing more than 100 individuals: Kaziranga National Park in Assam, India, has about 1,200 individuals, and Chitwan holds more than 600 individuals. Smaller populations occur in eleven other reserves in Nepal, India, and Pakistan (table 1.1). A detailed account of the historical distribution and demography of past and present-day populations of greater one-horned rhinoceros and their current trajectory is the subject of chapter 5.

Causes of the Recent Decline in Rhinoceros Populations

The rapid demise of rhinoceros is a direct result of relentless poaching pressure and loss of habitat. Poachers have wiped out or greatly reduced many populations of all five species. The extensive illegal trade in rhinoceros horn has been well documented (Martin and Martin 1987), and I describe it in detail in the next chapter.

If by some miracle poaching ceased in Africa, rhinoceros populations would rise rapidly because extensive habitat still remains to reestablish the species throughout much of its former range. Asia presents a different scenario. No habitat blocks remain in Asia that contain the equivalent to the Selous Game Reserve of southern Tanzania, an area of roughly 55,000 km², constituting a potential megareserve for black rhinoceros. In Asia rapid habitat loss, population growth adjacent to protected areas, and even human encroachment into protected areas threaten the rhinoceros. Lowland tropical forests (300 m above sea level) are critical habitats for Sumatran and Javan rhinoceros but are suffering from clear-cutting and conversion to agriculture or rubber and oil-palm plantations. Opportunities for conservation of rhinoceros in large tracts of lowland forests are possible only in the transboundary area of northeastern Cambodia, southern Laos, and the neighboring block in Vietnam. Lowland rain forests on the island of Sumatra—the most important habitats for Sumatran rhinoceros—have virtually disappeared. Lowland forests across Asia are under great threat from logging, and it is likely that by 2010 most of the forest remaining will consist of an archipelago of upland areas too difficult to log economically (Dinerstein et al. 1997; Wikramanayake et al. 2002).

The growth of human populations poses a much greater threat to habitat conservation in Asia than in Africa. The most preferred habitats of greater one-horned rhinoceros and Javan rhinoceros are the floodplains that double as sites of intensive cultivation. Remote areas in Africa that support rhinoceros populations are often devoid of settlements or are sparsely populated around the perimeter of conservation areas. Few such underpopulated areas exist within the rhinoceros range in Asia, with the exception of central Borneo.

Finally, wars, insurgencies, and the proliferation of automatic weapons in rural areas have taken their toll on rhinoceros. Hostilities in Indochina nearly decimated the remaining Javan rhinoceros in this region. Countries exporting arms to support civil wars and insurrections in Africa shoulder great responsibility for the demise of rhinoceros there. Traditionally, pit traps and spears were the techniques that Asians used most commonly used to kill rhinoceros. These were rather inefficient and even dangerous to the would-be hunter/poacher. If Asian and African governments disarmed their rural peoples, so that they could hunt large mammals using only traditional means, rhinoceros populations would be much more stable.

In Nepal, few private individuals have access to sophisticated firearms. Had Nepalese or Assamese villagers possessed the kind of arsenals at the disposal of Somali poachers, the populations of greater one-horned rhinoceros probably would not be at their present levels, despite strict protection by the Nepalese Army in core reserves. A dismal test of this hypothesis may have already occurred in the Manas tiger reserve of Assam, India. Here, an uprising by the heavily armed Bodo people virtually wiped out the existing greater one-horned rhinoceros population, once estimated at about 80 (Foose and van Strien 1997). India, which has strong laws governing the possession of powerful firearms, has reported poaching of greater one-horned rhinoceros as high as 450 individuals in the ten-year interval between 1986 and 1995. During the same interval, Nepal reported only 50 incidents (Menon 1996). Nepal has had poaching well under control since 1993 in Chitwan, through the joint activities of the Nepalese Army stationed inside the park and an active antipoaching information network operating outside the park.

CHAPTER 2

CULTURE, CONSERVATION, AND THE DEMAND
FOR RHINOCEROS HORN

I would take the last horn off the last living rhinoceros if it meant saving
my daughter from a fever that could be treated by nothing else.
—Chinese-American conservationist

A CONVENTION OF INDIAN GRIFFON VULTURES perched
in the treetops at the edge of the Rapti River in Chitwan National
Park, Nepal. Clearly, they were hoping for a feast of rhinoceros, but
they would have no such luck this time. For the moment, I was first
in the pecking order, allowed by park officials to measure the dead
male rhinoceros — which we had identified as M039 — that was
sprawled on the floodplain. Meanwhile, they conducted the official
inquest. This animal had died from wounds suffered in a fight with
the resident male. Thankfully, the horn was still attached when an
elephant driver reported the death to us. When a rhinoceros dies in
Chitwan, whether of natural causes or not, the senior park warden
must conduct a formal inquiry (figure 2.1). Nepalese officials say
that the procedure for filing a report on a dead rhinoceros is far
more involved than that required for a human being. I suppose this
is so because Nepal has many fewer rhinoceros than people, and rhinoceros are the official property of the king; the dead rhino was regarded as the loss of something priceless.

Hundreds of villagers encircled our small team of field hands

FIGURE 2.1 During an official inquest, rhino guards inspect a male rhinoceros that died from wounds suffered in a fight with another male. (Photo by Eric Dinerstein)

and park officials, their impatience growing steadily. If we kept them waiting much longer, the crowd would grow so large that each would receive only a small portion when the carcass was divided among them (figure 2.2). After the park guards removed the animal's horn, feet, and chest skin — to be stored in a government warehouse as the property of the king—we were told to step quickly away. An official gave the signal, and the Nepalese villagers rushed at the carcass and began slashing with traditional blades known as *kukhris*. Two hours later, the villagers had carved up and carted off the 2,000-kg rhinoceros, leaving barely a trace for the vultures to pick over.

One of my elephant drivers told me in a solemn voice that Nepalese consider every part of the rhinoceros valuable. This is an understatement. I have seen elephant drivers risk life and limb to collect fresh rhinoceros urine before the urine could soak into the ground. Rhinoceros urine has a pungent smell and an awful taste, at least to my uneducated palate. But many Nepalese and Indians believe that taken internally, it is an effective treatment for asthma and tuberculosis; they apply it topically to treat inner-ear infections. Rhinoceros urine is also rather expensive for the average villager, typically selling for nearly $1 per liter (all dollar amounts are expressed in U.S. dollars).

FIGURE 2.2 Local villagers wait to gain access to the carcass of an old male. (Photo by Eric Dinerstein)

The demand for rhinoceros urine is not the reason for its global demise; obtaining it does not involve killing the animal. The horn is the primary cause of the rhinoceros's decline.

The Trade in Rhinoceros Horn

As most people know, the rhinoceros horn is not at all like the horns of cattle, sheep, and goats, which contain a bony core. Rather, rhinoceros horn is composed entirely of densely appressed hairs. According to popular belief, Asian cultures value rhinoceros horn as a powerful aphrodisiac (figure 2.3). In fact, traditional Chinese medicine never has used rhinoceros horn as an aphrodisiac; this is a myth of the Western media and in some parts of Asia is viewed as a kind of anti-Chinese hysteria.*

---

*Another false assumption is that much of the pharmacopoeia of traditional Chinese medicine is animal based, overflowing with prescriptions that contain the diced, dried, and powdered body parts of dozens of endangered species. In fact, more than 80% of traditional Chinese medicines are plant based.

"It's supposed to be some kind of aphrodisiac, but it hasn't done jack for me."

FIGURE 2.3   The New Yorker Collection 1999. Mick Stevens from cartoonbank.com.

The reality is that traditional Chinese medicine considers rhinoceros horn an essential medicine. Its primary role is to alleviate life-threatening fevers for which no other cure exists. Thus rhinoceros horn is more highly prized than another controversial animal product, tiger bone, whose primary use is to treat rheumatism. Practitioners point out that nobody will die without access to tiger bone because rheumatism, while debilitating, is not a fatal condition. But some people will die, traditional Chinese medicine specialists say, because of the ban on using rhinoceros horn as a medicine. A short anecdote further clarifies the deep-rooted belief among Asians of the power of powdered horn as a fever depressant. A colleague of mine is acquainted with a Chinese-American who researches traditional Chinese medicine and is involved in the conservation effort. Nevertheless, this researcher stated unequivocally that he would take the last horn off the last living rhinoceros if it meant saving his daughter from a fever that could be treated by nothing else. We can therefore begin to understand the awe with which some cultures regard this strange appendage possessed by the five extant species of rhinoceros.

Despite the medicinal value of rhinoceros horn (studies have shown it to be an effective fever reducer), the main consumers — China, South Korea, Taiwan, and British Hong Kong — banned the manufacture and trade in rhino horn and its medicinal derivatives by the end of 1993. Before these governments implemented the 1993 ban, companies in East Asia manufactured millions of medicinal items each year that used minute amounts of horn. This large-scale production is now shut down in response to great pressure from conservation groups and governments in industrialized nations. North Korea is the only country that may still allow legal manufacturing of medicinal items from rhinoceros horn. However, illegal trade continues, on a small scale for medicines and perhaps a larger scale for speculation.

The Chinese government allows the use of confiscated rhinoceros horn medicines, which were locked up after the ban, as a fever depressant in hospital emergency cases. However, traditional Chinese medicine specialists in South Korea and Hong Kong calculate that one rhinoceros horn per consumer country per year (probably no more than ten such nations) would provide more than enough powdered horn to meet emergency needs. Thus the misconception that horn is still in high demand by 1 billion Chinese is dangerous. Probably because of this misconception, poachers and traders are still persecuting rhinoceros and stockpiling horns.

Market surveys conducted by conservationists working undercover unearthed three important findings. First, they showed that the horns of the three Asian species are about three times more expensive than those of African species; Asian rhinoceros horn fetches $30,000 per kilogram, whereas African horns are worth about $10,000 per kilogram. No published data are available on the sample size used to determine prices or how black market prices fluctuate. If the large discrepancy between the price of Asian and African horn is valid, it may in part reflect the historical demography of the five species — until quite recently, the African species were much more abundant than the Asian species. More likely, it is attributable to the belief among users of ground rhinoceros horn that species closer to home yield the most potent medicine.

In Yemen, a major smuggling destination, the major use of rhinoceros horn is not for traditional medicine but purely for ornamentation — as dagger handles for the traditional *jambiyas*.* In Yemeni society, the quality of the *jambiya* a man wears is symbolic of his status in the hierarchy, and men

---

*For information about the *jambiya* trade, I relied heavily on Martin, Vigne, and Allan (1997).

covet *jambiya* made from rhinoceros horn. Dagger handles may even be a much bigger problem than traditional Chinese medicine, as the known demand is larger and the main consuming country has not yet banned domestic trade. Since 1993, Yemen has assumed the dubious role of being the leading importing nation of rhinoceros horn. Martin, Vigne, and Allan (1997) estimate that since 1970 at least 67,000 kg of horn were imported to Yemen. The demand for rhinoceros horn since 1970 has been the leading cause of the demise of rhinoceros throughout vast areas of Africa. Assuming that average horn mass is 3 kg, the number of rhinoceros killed to supply this horn may be at least 22,350 animals. Consumption of horn dropped during the economic crisis in Yemen in the early 1990s but has since resumed. It seems that as income levels rise, more horn is consumed for *jambiyas*.

Reducing or Eliminating the Trade in Rhinoceros Horn

The ban on rhinoceros horn in traditional Chinese medicine has been effective, but perhaps even more important was establishing a working relationship with traditional Chinese medicine practitioners. In the early 1990s, I heard cynics declare that it would take generations to achieve a ban on the use of horn and that substitutes could not be found. Since then, water buffalo horn has replaced rhinoceros horn in the official pharmacopoeia of the People's Republic of China. This designation follows years of clinical studies and agreement by China's huge traditional Chinese medicine industry. In this era of new drug therapies, it is not farfetched to envision a time when high fevers will be treated with a variety of more effective drugs. The rhinoceros horn will no longer be considered an essential ingredient, even in those rare life-threatening cases when no other cure now is thought to exist.

The dagger handle trade remains a more vexing problem. Yet in many industrialized countries, formerly fashionable items — beaver fur, egret plumes, and, more recently, shatoosh wool from Tibetan antelopes — have rapidly become fashion treason once people understand the threat to the species involved. Clearly, the first step is to strongly urge Yemen to join the Convention on International Trade in Endangered Species (CITES) and to stop the import and domestic trade of rhinoceros horn — at the risk of sanctions from industrialized countries. Second, the CITES signatories must enforce existing laws that forbid trade in raw horn. If all those countries strictly enforced their laws pertaining to endangered species and their habitats, many fewer species would be endangered. Ultimately, I believe, the solution will involve strict enforcement, working with religious and secular leaders to promote the ban on killing rhinoceros, as decreed by the grand mufti of Yemen, and an overall awareness campaign. Promoting another style of *jambiya*, using precious and semiprecious stones, should be highly encouraged. World-renowned jewelers with a conservation commitment, such as Bulgari, could help in this transformation. Perhaps this is also where advertising expertise in industrialized nations, working with a Yemeni task force, could help promote product substitution and conservation. All these approaches are worth a try to reduce pressures on beleaguered rhinoceros populations.

Meanwhile, some wildlife officials in South Africa have been communicating with traditional Chinese medicine specialists in South Korea about the possibility of reopening limited bilateral trade in rhinoceros horn strictly for controlled medicinal use. While this is an attractive proposition to some suppliers and consumers, the governments that banned the trade are unlikely to lift their bans in the near future. CITES authorities in Hong Kong say that even limited trade would send the wrong message to the traditional Chinese medicine community about the still precarious status of rhinoceros in the wild and would pose a regulatory nightmare for those trying to stop illegal trade. More important, the bans have eased pressure on wild rhinoceros, allowing some populations to make a significant comeback. Given the efficacy of the bans, the availability of substitute medicines, and the risks of stimulating illegal trade posed by limited legal trade, most parties agree that the use of rhinoceros horn as medicine should remain a thing of the past.

CHAPTER 3

VANISHING LANDSCAPES:
THE FLOODPLAIN ECOSYSTEM OF THE TERAI

Even the survivors of the current extinction crisis will face an
unprecedented situation; their evolutionary capability will be greatly
diminished. The reason is that evolution in large animals is impossible
without a lot of space.
—Michael E. Soulé, "The End of Evolution?"

THE THUNDER OF THE RIVER woke me from a fitful sleep.
Torrents of rain had engulfed the Chitwan Valley for three solid
days, and now in September, the last month of the monsoon, the
earth could hold no more. I rushed down to the banks of the Rapti
River and joined villagers who stood gaping at the brown floodwaters surging toward India that were sweeping along uprooted trees
as if they were toothpicks. The scene was almost biblical or, more
appropriately, something from the Hindu scriptures, in the display
of the raw power of nature. Yet this scene plays out every year during the monsoon. Each year, the rivers of Nepal's Terai zone—the
low-lying strip of land at the base of the Himalayas—breach their
channels and change their course. The only difference from year to
year is the severity of the floods. Farmers face ruin if the overflowing rivers bury the fields in sand. Conversely, they will reap a rich
harvest if the river deposits its silt load and recharges croplands
with nutrients washed down from the hills.

For the terrestrial wildlife of the Terai, the flood season is time
to seek higher ground or be swept away. In this vanishing ecosystem, the annual monsoon floods are the most striking component

of the regional ecology and a fundamental influence on the structure of animal and plant communities. As the floodwaters recede, the silt deposited in the grasslands becomes a thick layer of fertilizer. Tall grasses respond to this treatment and virtually erupt from the silt. For grazers it is a time to feast on new shoots. For small ungulates, such as hog deer, that require the cover of elephant grasses to escape from predators, this is a time of great uncertainty. But the native species of the Terai have adapted to this dramatic regime of disturbance. Species that failed to cope with this annual, predictable, and at times nearly catastrophic disturbance would have been selected out of the flora and fauna long ago. Unfortunately, few places are left where the cycle of the floods and the annual recovery of wild landscapes still plays out.

Chitwan's floodplain and terraces produce what is arguably called the world's tallest grasslands. "Elephant grasses" (figure 3.1) reach 6 to 8 m by the end of the monsoon, usually in October. This biologically productive habitat supports tigers, elephants, and other endangered large mammal species, as well as the rhino. However, this same biological productivity translates into high agricultural fertility; therefore, much of the grasslands has been converted to agriculture.

FIGURE 3.1 *Saccharum spontaneum* grassland (elephant grass) forms the first stage of succession in recently flooded areas. (Courtesy John Lehmkuhl)

The Global Significance of the Terai–Duar Savannas and Grasslands

A recently published global analysis by the World Wildlife Fund–United States of priorities for conserving biodiversity identified the Terai–Duar savannas and grasslands as one of the important ecoregions because this area is one of the most biologically outstanding grasslands on Earth (Olson and Dinerstein 1998) (box 3.1). These grasslands fringe the base of the outermost foothills of the Himalayas, extending from Dehra Dun in Uttar Pradesh, India, across the Nepalese Terai zone to the Duar grasslands of Bhutan (figure 3.2). One feature that elevates the Terai–Duar savannas to Global 200 status is the diversity of its ungulate species and extremely high levels of ungulate biomass recorded in riverine grasslands and grassland–forest mosaics (Dinerstein 1980; Seidensticker 1976) (table 3.1). Many Terai grasslands and floodplain forests support five species of deer — swamp deer, sambar, axis deer, hog deer, and barking deer — an unusually diverse assemblage of cervids for any given site (figure 3.3). Four very large herbivores — Asian elephant, greater one-horned rhinoceros, Indian bison (seasonal occupants), and nilgai, or blue bull (in drier grasslands) — all coexist here.* Several endangered mammalian herbivores, including the Asiatic wild buffalo and a small herbivore restricted to tallgrass, the hispid hare (Chapman and Flux 1990), now have very restricted ranges. Pygmy hog are another highly endangered species of the tall grasslands (Oliver 1980). More impressive than the ungulate diversity is the extraordinary level of ungulate biomass. Seidensticker (1976) portrays the ability of Chitwan's grasslands to support levels of ungulate biomass (wild and domestic) that exceed all other sites in Asia and rival some in East Africa (Eisenberg and Seidensticker 1976; Seidensticker 1976).

Throughout the world, tall grasslands such as those in Chitwan are both rarer and more threatened than short grasslands. Tall grasslands are indicators of mesic, or wet, conditions and nutrient-rich soils — attributes prized by farmers. Less than 1% of the tallgrass prairie of the United States

---

*The extirpation of swamp deer and wild water buffalo from Chitwan in the 1950s has diminished its assemblage of large ruminants. Swamp deer are still abundant, however, in other Terai reserves (Dudhwa National Park, India, and Royal Sukla Phanta Wildlife Reserve, Nepal), and water buffalo hold on in the Koshi Tappu Wildlife Reserve in eastern Nepal. Mishra and Jeffries (1991) provide a nearly complete list of Chitwan's mammals, except for Chiroptera (bats).

The conversion and degradation of the world's temperate, subtropical, and tropical grasslands have largely gone unnoticed as international conservation agencies concentrate on tropical rain forests. The Global 200 analysis by the World
Wildlife Fund specifically recognizes the significance of these grassland and savanna ecoregions (figure B3.1) by taking a representation approach in ranking
the most outstanding areas for biodiversity on land, in freshwater, and in the
marine environment. The analysis achieved representation by dividing the
world first into continental realms and then into biomes such as grasslands,
tropical moist forests, deserts, lakes, large rivers, coral reefs, and mangroves. For
each biogeographic realm or ocean basin, researchers selected the most biologically important ecoregions that occur within a given biome. In this particular
context, an ecoregion is simply an ecosystem of regional extent. More specifically, it is a large unit of land or water characterized by a geographically distinct
set of characteristic species, habitats, dynamics, and environmental conditions.

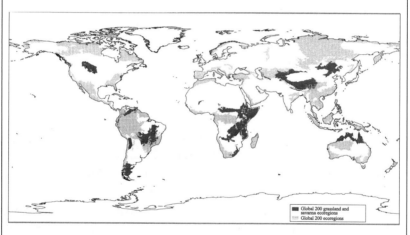

FIGURE B3.1   Global 200 grassland and savanna ecoregions. (Adapted from Olson and
Dinerstein 1998:510)

Because most of the world's grasslands have been converted to agriculture or
overgrazed by domestic livestock, the loss of intact grassland habitats has led to
the endangerment of grassland-associated fauna around the world. The most
endangered bird species in continental North America are not the neotropical
migratory songbirds of the Appalachian forests but bobolinks, dickcissels, and
other birds associated with the tall- and shortgrass prairies. The assemblage of

BOX CONTINUED

bison, black-tailed prairie dogs, black-footed ferrets, wolves, and grizzly bears has disappeared from the most extensive temperate grassland system in the world, the northern High Plains. In South America, Patagonian grasslands have little room for wild camelids, such as guanacos, now that sheep ranching is the dominant economic force.

As the world's intact grasslands disappear, vital ecological phenomena, such as spectacular ungulate migrations, are rapidly declining. Conversion of the steppe grasslands and intense poaching have greatly affected the migrations of saiga antelope and endangered them in Central Asia. American bison herds are now considered an exotic form of livestock, raised on scattered ranches throughout the United States. Only in the *Acacia* savannas of East Africa, with their extensive wildebeest migrations, and on the Tibetan plateau, with the migrations of kiang (Tibetan wild ass) and Tibetan antelope, do waves of large herbivores still roam about the tropical and temperate latitudes. The presence of these important and increasingly rare ecological phenomena warrants inclusion of all these phenomena in the Global 200 analysis. Although grasslands may be lower in species richness and endemism than tropical rain forests, the loss of the adaptations that grassland-dwelling species evolved over long time periods would be an incalculable blow to global biodiversity.

and Canada remains intact; virtually all of it has been converted to row-crop agriculture, and most remnants are postage stamp–size parcels of only a few hectares. The Terai–Duar savannas and grasslands are not much different. Perhaps no more than 2% of the alluvial grasslands of the Indo-Gangetic Plain remains intact, and the best-conserved examples of flood-plain grasslands are in Chitwan, Sukla Phanta Wildlife Reserve, Manas Wildlife Sanctuary, Dudhwa National Park, and, to a lesser extent, Royal Bardia National Park (figure 3.2).

The Parsa Wildlife Reserve east of Chitwan, and the adjacent Valmiki tiger sanctuary on the border with India, form an important transboundary area for wildlife conservation in general and for tigers in particular (figure 3.4). This tiger conservation unit (chapter 11) was deemed the most important of all the alluvial grassland units containing tigers on the Indian subcontinent (Dinerstein et al. 1997; Wikramanayake et al. 1998a). The tiger conservation unit supports a healthy leopard population and at least a small population of the rare clouded leopard, the last new large mammal to be reported for Chitwan; the clouded leopard's appearance in Chitwan rep-

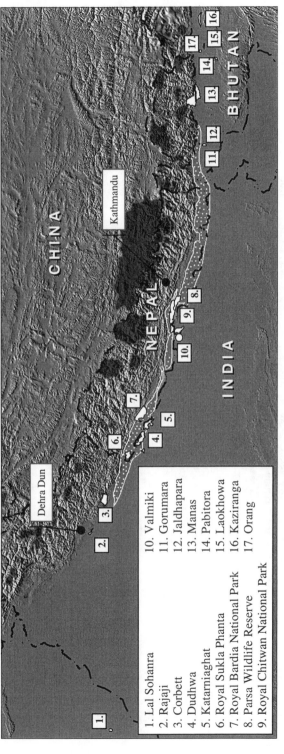

FIGURE 3.2 Major protected areas of the Himalayas (*dark gray*). This digital elevation model base map shows in white boxes the Terai wildlife reserves and national parks along the Indo-Gangetic Plain and the reserves along the Brahmaputra River Valley (for example, Kaziranga). The white dotted area is the Duar grassland ecosystem. (Courtesy Tom Allnutt, Conservation Science Program, World Wildlife Fund)

TABLE 3.1. CRUDE BIOMASS ESTIMATES (KG/KM^2) FOR UNGULATES IN INDIA, INDONESIA (JAVA), NEPAL, AND SRI LANKA

	Dry thorn forest and steppe	Dry semide-ciduous forest with scrub	Deciduous to semievergreen forest				Reference
			Moist semi-deciduous forest	Intermediate	Gallery forest and alluvial plain	Tropical evergreen forest with meadows	
Wild mammalian biomass							
Bharatpur (Rajastan, India)	1,617	—	—	—	—	—	Spillett 1967a
Gir (Gujarat, India)	—	383	—	—	—	—	Berwick 1974
Wilpattu (Sri Lanka)	—	766	—	—	—	—	Eisenberg and Lockhart 1972
Gal Oya (Sri Lanka)	—	886	—	—	—	—	McKay 1973
Kanha (Madhya Pradesh, India)	—	—	1,708	—	—	—	Schaller 1967
Chitwan (Nepal)	—	—	—	1,790	—	—	Seidensticker 1976
Jaldhapara (West Bengal, India)	—	—	—	—	984	—	Spillett 1967b
Kaziranga (Assam, India)	—	—	—	—	2,858	—	Spillett 1967b
Ujung Kulon (Java, Indonesia)	—	—	—	—	—	492	Hoogerwerf 1970
Domestic ungulate biomass							
Gir (Gujarat, India)	—	6,171	—	—	—	—	Berwick 1974
Kanha (Madhya Pradesh, India)	—	—	4,678	—	—	—	Schaller 1967
Chitwan (Nepal)	—	—	—	28,076*	—	—	Seidensticker 1976

* Maximum density.

SOURCE: Adapted from Eisenberg and Seidensticker (1976:300, table 4).

A

B

FIGURE 3.3 Royal Chitwan National Park is home to five species of deer. (A) Also known as the spotted deer, or chital, the axis deer is the most abundant ungulate in the park. (B) The largest deer in Chitwan is the sambar. Swamp deer were common in Chitwan until the mid-1950s, when they were extirpated. (Courtesy Colby Loucks/WWF)

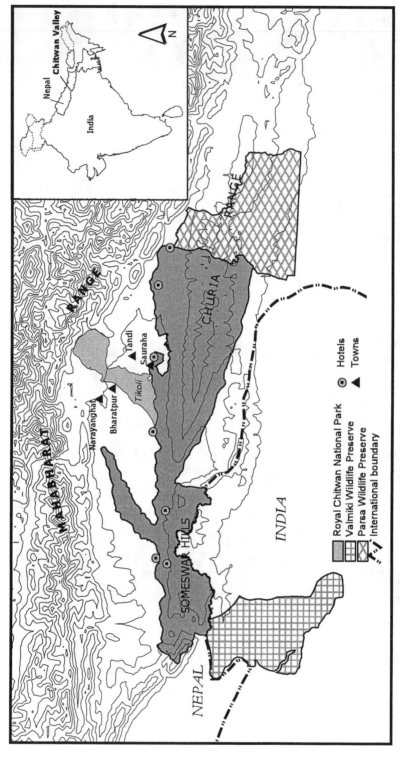

FIGURE 3.4 The Chitwan-Parsa-Valmiki Tiger Conservation Unit. (Courtesy Wesley Wettengel, Conservation Science Program, World Wildlife Fund)

resents a range extension of more than 250 km to the west of its former territorial boundary (Dinerstein and Mehta 1989).

In addition to its high ungulate and carnivore diversity, Chitwan is home to 524 bird species (Herm Baral, personal communication, 1999). This is one of the highest recorded levels in Asia or for any national park and represents almost 6% of the world's known birds. Nepal's list of threatened birds, 130 breeding and wintering species, includes 14 species that are grassland specialists. Many of these grassland-associated birds were once common in the Chitwan Valley: Bengal florican, lesser florican, Sarus crane, common crane, and large grass warbler. Nowadays the numbers of these birds are declining steadily. Riverine grasslands in the Terai also provide critical habitat for endangered reptiles, including the rare primitive crocodilian, the gavial, or gharial (Maskey 1979); mugger crocodile; and softshelled turtles (Zug and Mitchell 1995).

The Biological Setting

The Chitwan Valley is located in south-central Nepal at 120 to 200 m above sea level. Royal Chitwan National Park lies essentially south of the Rapti River and covers 932 km², while most of its buffer zone is on the north bank. The most recent demarcation of the buffer zone (in 1998) expanded the area under park management by another 750 km².

Chitwan has three distinct seasons. The monsoon season is perhaps the most dramatic and important in terms of ecosystem dynamics. Chitwan averages about 2,400 mm of rainfall per year, 90% of which falls during the four-month monsoon season (June–September). Running across the Terai is a distinct east–west rainfall gradient, so Chitwan receives more rainfall annually than do Bardia and Sukla Phanta in far western Nepal. November through mid-February marks the cool season, when temperatures can drop to 9°C. Mid-February through May is the hot season; high temperatures in April and May regularly reach 36°C. Premonsoon showers and severe windstorms are a regular feature in the late hot season. In chapter 6, I describe climate patterns in greater detail as they relate to wallowing behavior in greater one-horned rhinoceros.

Ecosystem Dynamics

Large herbivores and carnivores thrive on Chitwan's floodplain for many reasons. First, the floodplain sits at the base of the Himalayas, the world's youngest and highest mountain chain. The steepness of the Himalayan terrain and the fragility of the soils combine with the monsoon season's high rainfall to cause frequent and severe floods and tremendous rates of soil erosion. Significant amounts of topsoil wash into India and the Bay of Bengal. In fact, erosion is such a serious problem that some observers have suggested that Nepal's greatest export is its topsoil. Nevertheless, every year the main rivers that flow through Chitwan—the Narayani, the Rapti, and, to a lesser extent, the Reu—pound across the floodplain, burying grasslands in more than 1 m of silt (figure 3.5a). The most severe floods carry such large deposits of sand that agricultural lands are virtually unusable for several years thereafter. Such was the case during the monsoon of 1993 when a severe flood forced thousands of people to evacuate the Chitwan Valley. Areas buried in silt returned to tall grasslands by the end of the following monsoon season. The annual load of nutrients recharges the low-lying areas, which are inundated for only a few days, promoting unusually high rice yields. Thus the same phenomenon that supports an astounding human population density along the floodplains of the Brahmaputra and Ganges Rivers in Bangladesh and India also supports the high levels of wild ungulate biomass recorded in these habitats.

Annual fires also play a major role in shaping the landscape of the Terai (figure 3.5b). Before humans inhabited the Chitwan Valley, lightning strikes started fires. However, a greater proportion of Chitwan's grasslands and forests probably are burned today as a result of human activities than were burned in prehistoric times. Fire and, especially, floods are common annual disturbances. Although locally severe, these events, especially floods, are a major structuring force in this ecosystem.

The high productivity of the riverine systems of the Terai may also be responsible for notable characteristics of the local tiger population. For example, Sunquist, Karanth, and Sunquist (1999) have noted that tigers in Chitwan have the smallest home ranges and highest densities in Asia. Male tigers (figure 3.6) captured in Chitwan are also the largest free-ranging *Panthera tigris* captured to date. Both a male captured by Smithsonian and Nepalese researchers in 1980 (tiger M105) and another captured by our project in 1984 (M026) exceeded 270 kg. Amur tigers of the Russian Far East are reported to be the largest in body mass among tiger populations spread across Asia. However, no male Amur tigers captured to date have exceeded the records of body mass reported for Chitwan (Dale Miquelle, personal communication, 1999). Comparisons of grassland productivity with body mass

A

B

FIGURE 3.5 Dynamic processes in the Terai ecosystem include (A) annual floods during the monsoon season and (B) frequent grassland fires set by lightning strikes and people. (Courtesy John Lehmkuhl)

FIGURE 3.6 A sedated adult male tiger. Tigers are the largest predator of the Terai. (Photo by Eric Dinerstein)

for other large predators native to Chitwan (leopards, jungle cats) and widely distributed in Asia would be interesting and would help support this theory.

Vegetation Patterns

Proximity to a major river, duration of inundation during the monsoon season, elevation, and soil conditions largely determine the dominance of particular vegetation types and successional patterns on the Rapti River floodplain in Chitwan and in other parts of the Terai grasslands (Dinerstein 1975, 1979, 1985). The effect of human-caused fires during the dry season and overgrazing by domestic livestock typically is to simplify the shift of vegetation associations to earlier successional stages of development.* Composition and structure of riverine forests, grasslands, and savannas both determine and are highly influenced by the feeding behavior of giant herbivores.

*For greater detail on the vegetation of the Rapti River floodplain, see Dinerstein (1975) and Lehmkuhl (1989). Vegetation analyses of forest stands and grassland associations in Royal Bardia National Park — which I mention frequently in this book as the site of translocations — can be found in Dinerstein (1979).

Saccharum spontaneum (kans, in Nepali) is the first type of grassland to colonize the exposed silt plains after the retreat of monsoon floods. This tallgrass species (figure 3.1) often occurs in almost pure stands, forming a thin strip on the first terrace of the floodplain, which can range from less than 100 m to more than 1 km in width. Without question, this is the key habitat for rhinoceros and other large mammals. Saccharum spontaneum is the most nutritious of the abundant tallgrass species in the park, and rhinoceros seek it out. Although grasslands account for about 15% of the park's total area, only a small fraction, perhaps less than 5% of the park, consists of this grassland type. Both wind and water, and perhaps large mammals, disperse S. spontaneum. In September, domesticated elephants emerge from stands of S. spontaneum with the white inflorescences (flowers) clinging to the hair on their faces and foreheads. Grazing lawns (chaurs) are most commonly found within S. spontaneum grasslands (McNaughton 1984). They consist of very short swards (5 cm) of mixed grasses maintained by intense grazing by greater one-horned rhinoceros (figure 3.7) and other large ungulates. Imperata cylindrica, Chrysopogon aciculatus (kuro), Eragrostis spp., and many shortgrasses typically dominate these swards.

Saccharum benghalensis (baruwa) grasslands form the terrace just

FIGURE 3.7 Intense grazing by rhinoceros and other ungulates turns grasslands into short-cropped lawns. (Courtesy Colby Loucks)

above the *S. spontaneum* band along the rivers of the Terai. Greater one-horned rhinoceros feed on *S. benghalensis* in the shoot stage. *Saccharum benghalensis* often dominates the vegetative cover of rhinoceros latrines in *S. spontaneum*–dominated grasslands and can be recognized from afar by a conspicuous growth form (bent leaf blades at the tip) and greenish blue leaves.

Imperata cylindrica (*shiru*) grasslands (figure 3.8) cover old village sites within the park and expand where grasslands are burned repeatedly. *Imperata* is a valuable thatch grass but is largely avoided by greater one-horned rhinoceros and other ungulates when it has grown past the shoot stage. *Imperata* is one of the few grasses that contains secondary compounds, a feature found in many dicotyledonous plants to presumably thwart herbivores. This may explain its low palatability in mature stages because many plants increase the amount of secondary compounds in leaf tissues upon maturation.

Greater one-horned rhinoceros favor several other grasses. *Cymbopogon* spp. (*ganaune gans*) is another relatively short grass species that occurs in distinct associations on the floodplain and is eaten by both rhino and elephant. The tallgrasses *Arundo donax* and *Phragmites karka* (*narkot*)

FIGURE 3.8   An axis doe and her fawn traverse grassland dominated by *Imperata cylindrica*. This species can persist for decades on disturbed sites. (Photo by Eric Dinerstein)

surround oxbows and lakes. Greater one-horned rhinoceros and domesticated elephants seek them out as well.

Grasslands dominated by the tallest grass of Chitwan, *Narenga porphyracoma* (*phank*), attract greater one-horned rhinoceros for only a few weeks in spring when new shoots emerge after fires. This association is by far the most dominant in Chitwan (Lehmkuhl 1989), but all ungulates except Indian bison largely avoid *phank*. Savannas of *N. porphyracoma* with large *Bombax ceiba* trees and understories of *Narenga* and much smaller trees such as *Xeromphis uliginosa* (*pidar*) are common.

The least attractive grasslands are those dominated by *Themeda arundinacea* (*ooreli*) in monodominant stands or mixed with *Narenga*. *Themeda* grasslands are most common at the edge of *sal* forests and grasslands. Both elephants and greater one-horned rhinoceros avoid *Themeda*. Rhinos find *Themeda villosa*, a species of the mixed grasslands, more palatable.

The park contains several types of riverine forest associations. In the eastern section, the most common association is of the dioecious *Trewia nudiflora* (*bhelur*) and *Bombax ceiba* (*simal*) (figure 3.9). *Trewia* can account for more than 50% of all individual trees in riverine forest patches, followed

FIGURE 3.9 *Trewia nudiflora* (Euphorbiaceae), which is flushing new leaves, and *Bombax ceiba* (Bombacaceae), the tallest trees with spreading crowns, which displays bright red flowers, dominate this riverine forest. (Photo by Eric Dinerstein)

by *Ehretia elliptica* (*dhatrung*) (Dinerstein 1992). Greater one-horned rhinoceros heavily browse the understory shrubs *Coffea benghalensis* (*baramasi*) and *Murraya koenigii* (*laathikat, asare*) in the cool season. They seek *Trewia* fruits in the summer but avoid most of the common shrubs, *Colebrookia oppositifolia* (*dhurshil*), *Pogostemon benghalensis* (*rudilo*), and *Clerodendron viscosum* (*bhanti*). *Albizia lucida* (*parke*), a species eaten by greater one-horned rhinoceros, dominates another type of riverine forest. This forest has a dense understory of *Myrsine chisia* (*bhilaune*), which the rhino avoids. Where the forest has been highly degraded and its canopy cover removed, the stoutly thorned climbers *Acacia rugata* (*areli kara*) and *Mesoneuron cuculata* (*bokshi kara*) are common. Greater one-horned rhinoceros will seek out the legumes of *A. rugata*.

On the floodplains of the Narayani, pure or mixed stands of *Dalbergia sissoo* (*sissoo*) or *Acacia catechu* (*khair*) are common. Greater one-horned rhinoceros heavily browse the seedlings of both. *Persea duthii* (*kaulo*) and the sought-after browse of *Murraya paniculata* (*laathikat*) dominate other evergreen forests, while *Syzigium cumini* (*jamun*) dominates still another type of evergreen forest, and *Ficus racemosa* (*gular* or *dumri*) dominates the forest found along stream courses.

Along the edge of the buffer zone, scrub vegetation holds a particular attraction for greater one-horned rhinoceros. During the winter, they feed heavily on several browse species prevalent there, including *Callicarpa macrophylla* (*daikamala*) and *Cassia tora* (*saano taphre*). They even favor the stick-tight fruits and stems of the weedy *Xantium stromarium* (*bhedi kuro*, or cocklebur), an introduced weed thought to be dispersed on animal fur.

Croplands attract greater one-horned rhinoceros and other ungulates, and farmers must stand guard at night to scare animals away. The main local crops are rice, corn, wheat, lentils, and mustard. Greater one-horned rhinoceros are partial to rice, corn, and wheat at ripening. They seek out and devour hot chili plants but feed only sparingly on the mustard crop. Rhinos do not range far for their agricultural foraging. Damage to crops is restricted to 1 km from the park boundary (Jnawali 1995; Laurie 1978).

Shorea robusta (*sal*) is the dominant forest type of northern India and lowland Nepal (figure 3.10). Much of Royal Chitwan National Park, perhaps 70% of the area, and the adjoining Parsa and Valmiki Wildlife Reserves consist of pure or mixed stands of this vegetation type. *Sal* is a valuable timber tree, but large ungulates avoid it as forage. The high concentrations of tannins in *sal* leaves probably make them unpalatable to herbivores. The low

FIGURE 3.10 *Sal* forest covers most of Royal Chitwan National Park, especially upland areas. Visitors often see langurs (monkeys) in the branches of these trees. (Courtesy World Wildlife Fund)

palatability of *sal* forest species — such as *Terminalia alata* (*saj* or *asna*), *T. belerica* (*barro*), *T. chebula* (*harro*), *Lagerstromia parviflora* (*bhotdhairo*), *Dillenia pentagyna* (*tantari*), *Syzigium operculata* (*kyamun*), *Carya arborea* (*kumbi*), and *Buchanania latifolia* (*kalikat*) — further reduces this forest's attractiveness for many ungulates. Another unattractive feature of some *sal* forests is their lack of moisture. Hillside, or upland, *sal* forest, a subtype that covers much of Chitwan, sits on a geologic formation known as *bhabar,* which is composed of large boulders and coarse material. *Bhabar* soils have poor water retention capacity. Thus seasonal streams in the hilly tracts of Chitwan and the rest of the Terai hold water only during a monsoon season and for a short time thereafter. Perhaps the most important use of these upland *sal* forests by greater one-horned rhinoceros is as a refuge during floods.

Protection Efforts

Most accounts of the recent history of Nepal recognize two important periods: the 100-year rule of the Rana family — which controlled Nepal until the family's overthrow in the early 1950s — and the rise of the present monarchy marked by the ascent to the throne of King Tribhuvan. While these ruling dynasties have had a major influence on conservation efforts, other historical events are also worth considering.

The Eradication of Malaria

A game warden in Africa once said that the best agent of conservation in eastern and southern Africa was the tsetse fly. In sub-Saharan Africa, the threat of sleeping sickness restricted livestock to areas free of tsetse fly, allowing wild ungulates to roam over much of the continent. In the Terai belt, a similar situation prevailed because of malaria-bearing mosquitoes. Until the early 1950s, Chitwan and much of the Nepalese Terai were noted for a deadly strain of malaria. Accounts at the beginning of the nineteenth century suggest that the government of Nepal deliberately maintained a forested zone to function as a malarial barrier to discourage invasion from the south (Oldfield 1880 and Burkhill 1910, cited in Laurie 1978). The only residents of the Terai were the indigenous Tharu, who had somehow developed a degree of natural resistance to malaria. The ecological effect of the Tharu was negligible, because of their limited numbers and their custom of farming and herding at the edge of the forest rather than inside it (Giselle Krauskopf, personal communication, 1987).

The Chitwan Valley, covering approximately 2,600 km², remained essentially a primeval jungle of virgin *sal* forests, riverine forests, and tall grasslands. Early descriptions of the Chitwan Valley attest to the diversity and density of large mammals there. Elephants, rhinos, tigers, leopards, sloth bears, wild pigs, buffalo, Indian bison, and five species of deer were abundant (Oldfield 1880). During the two coldest winter months, when the threat of malaria abated, the Rana family and guests, often foreign royalty, staged grand hunts in Chitwan. Using hundreds of trained elephants for pursuit, these hunting expeditions killed large numbers of tigers and rhinos (figure 3.11). For example, 38 rhinos and 120 tigers were shot during the 1937/1938 season (Smythies 1942). Although excessive, current demographic data on greater one-horned rhinoceros and tigers suggest that these hunts made only a small dent in wild populations at that time. The 38 rhinos killed were probably fewer than the number of calves born in Chitwan at that time (extrapolating data from my own study of demographics). The high reproductive rates of tigers and the abundance of prey species led to

FIGURE 3.11 Royal hunts killed large numbers of rhinoceros and tigers.

their rapid recovery. Hunts also rotated among different areas, and large hunts were rare, giving local populations of tigers and rhinos time to recover. Furthermore, the hunting of rhinoceros was the exclusive privilege of the ruling family and its royal guests; poaching of rhinoceros brought the death penalty (Oldfield 1880).

Because of the absence of people and heavy hunting pressures, megaherbivores and carnivores remained abundant through much of the Rana era. Only the wild elephant population was in decline, compared with levels observed at the end of the nineteenth century. This population was probably migratory; losses accrued elsewhere may have affected the Chitwan population. Captures for domestication probably also reduced the wild elephant herd (Chris Wemmer, personal communication, 1988).

The Postmalaria Eradication Period

By the early 1950s, a number of factors had coincided to dramatically change the landscape and threaten the wildlife of the Terai. First and foremost was the fall of the Rana family and with it the Rana policy to discourage the clearing of the Terai forests. Still, threats to the wildlife habitats of Nepal were largely confined to some new areas settled by poor hill farmers. Then, beginning in the late 1950s, eradication programs—financed largely by foreign aid agencies—started to wipe out the threat of malaria. The control and eventual eradication of malaria opened up the Chitwan Valley to thousands of impoverished hill farmers. People from many different ethnic groups, many of whom were experienced hunters in the hills, moved into Chitwan Valley. The human population rose from 36,000 in 1950 to 100,000 in 1960. By 1959, the entire length of the valley was settled, and 70% of the forest and grassland habitat had been converted to agriculture.

As people spread across the Chitwan Valley, poaching of rhinos for their valuable horn increased, and the rhino population plummeted. Starting with an estimated 1,000 individuals, one observer estimated that there were 800 in 1950, Stracey (1957) suggested a further decline to 400 by 1957, and Gee (1959) put the number at 300 two years later. The precipitous decline led to the first step in the creation of Royal Chitwan National Park. In the late 1950s, Nepal established Mahendra Deer Park in the northwestern part of the valley. A wildlife sanctuary south of the Rapti River was also proposed for a ten-year trial period. Unfortunately, human settlements in both the deer park and the proposed sanctuary presented resistance. In response, the government moved almost all the villages inside the sanctuary to areas north of the Rapti River. This bold move was the single most important step in the creation of Royal Chitwan National Park. Without this measure, Chitwan would not have become the kind of sanctuary for the greater one-horned rhinoceros population that it is.

The area's human population continued to increase rapidly, through both in-migration and births. Poaching and habitat loss continued, but at a slower pace than before. Rhinoceros populations were still declining, and by the mid-1960s Spillet and Tamang (1967) were speculating that about 100 greater one-horned rhinoceros remained in Chitwan. Despite conservation measures, both wild water buffalo and swamp deer disappeared in the late 1950s. A number of unconfirmed sources attribute the loss of these populations to rinderpest, but no published accounts exist.

By 1971, the human population in the district had reached 185,000, and the conversion of the greater one-horned rhinoceros's preferred habitats continued inexorably. Using helicopter surveys, Graeme Caughley, working with Hemanta Mishra of the Department of National Parks, estimated the greater one-horned rhinoceros population at 81 to 108 individuals in June 1968 (Caughley 1969). A second helicopter survey in May 1972, followed by a ground survey using ten trained observers atop elephants, yielded an estimate of 120 to 147 animals (Pellink and Upreti 1972). Both Laurie's (1978) study and my efforts detail the labor-intensive nature of counting rhinoceros accurately and how easy it is to miss individuals. These early surveys probably underestimated the total number of rhinos in the park area. Between the mid-1960s and the mid-1970s, the population increased to between 270 and 310 individuals (Laurie 1978).

In 1973 the government of Nepal officially established the 544-km^2 Royal Chitwan National Park, and it encompassed the proposed wildlife sanctuary on the south bank of the Rapti River. Then the government and the Smithsonian Institution's National Zoological Park signed a formal agreement to conduct the first radiotelemetry studies of tigers. The agreement paved the way for eighteen years of field research on tigers, their prey, and greater one-horned rhinoceros. Many Nepalese and other graduate students were trained and conducted their field research in Chitwan, which soon became the best-studied wildlife reserve in Asia (Wemmer, Simons, and Mishra 1983). The King Mahendra Trust for Nature Conservation, supported by the Smithsonian, became Nepal's first conservation nongovernmental organization. The trust assumed leadership of the research program in 1984. As a result of research findings from the tiger project, the government extended the park's limits several times, up to its current size of 932 km². Chitwan, together with the additions of the Parsa Wildlife Reserve (499 km², established in 1984) to the east and the Valmiki Wildlife Reserve (336 km²) to the south (on the Indian side of the border), form the conservation complex known as the Chitwan-Parsa-Valmiki Tiger Conservation Unit (Dinerstein et al. 1997) (figure 3.4; chapter 11). These extensions are mainly important for tiger conservation; only a minor portion of them contains greater one-horned rhinoceros habitat because most of their territory is hillside *sal* forest, which is dry for much of the year.

In the early 1970s, concern about the increased poaching of rhinos inside the park spurred the reorganization of the protection force. In 1976 the Royal Nepalese Army replaced the rhino patrol, a poorly equipped but earnest group of forest guards. Today a division of army soldiers is based inside the park, and the rhino patrol covers the habitats outside the park boundary. But pressures outside the park were mounting (Sharma 1991). Buffer-zone areas that had healthy populations of greater one-horned rhinoceros in 1988 (Dinerstein and Price 1991) contained few or no animals in 1994 (Yonzon 1994).

The 1980s saw Chitwan become the major destination in Asia for nature tourists from industrialized countries; the decade also saw the spawning of a major privately run ecotourism industry (chapter 10).

Landscape-Scale Conservation and Local Guardianship of Endangered Species and Habitats

Throughout tropical Asia and Africa, it was becoming increasingly clear that strict protection measures, no matter how successful in the short run, are merely a temporary solution. Pressures on park habitats to meet local demands for firewood, fodder, and wild game continued to mount. More often than not, subsistence farmers work the land bordering reserves. In the 1960s, the biggest concern for these poor villagers was whether they could grow enough food to avoid starvation. While the "Green Revolution" has made the Indian subcontinent self-sufficient in rice, the challenge today is to find enough firewood to cook.

The demand for firewood and the intensity of human threats kept multiplying; by the late 1980s, the park, even with Parsa and Valmiki, was not big enough to hold viable populations of tigers (Kenny et al. 1995; Smith, McDougal, and Sunquist 1987) and greater one-horned rhinoceros (Dinerstein and Price 1991). Nearby villagers, lacking the basic resources for survival, were degrading the buffer zones rapidly and encroaching on the park. Habitat outside formally protected reserves badly needed zoning. But the involvement and support of local people were absolutely essential for the creation of viable buffer zones, wildlife corridors, and areas under the jurisdiction of agencies such as the Department of Forestry.

In the early 1980s, Nepalese conservationists began to lobby for laws that would allow local village committees to take over management of degraded Department of Forestry lands adjacent to protected reserves. Finally, in 1993 a major reform in national policy allowed the legal creation of buffer zones around existing protected areas. Local user group committees took over management of these zones; the law required that they develop effective management plans based on sustainable use of resources. Additional landmark legislation came in 1995, when parliament ratified a series of by-laws requiring that 50% of the revenue generated by protected areas be allocated to local development programs in these buffer zones instead of to the Ministry of Finance. Now operational, these two initiatives paved the way for establishing legal economic incentives to reduce pressures on core reserves and to conserve wildlife habitats outside parks. Most important, they allowed villagers to become partners in the recovery of the buffer zones and to serve as guardians of endangered wildlife and their habitats. The recovery strategies for Chitwan took full advantage of the new legislation. Without these policy changes, the new approaches to local guardianship of vanishing mammals and landscapes (described in part III) would have been impossible.

The Biology of an Endangered Megaherbivore

THE EMERGING SCIENCE OF conservation biology — the science of scarcity and diversity — has created a framework for assessing the demographic, genetic, and environmental risks facing small isolated populations of endangered species. These, along with landscape-scale issues such as habitat fragmentation, reserve size, core areas, connectivity of habitats, and maintenance of ecological processes, constitute the substance of research in this emerging field of science. Part II introduces some of these central issues of conservation biology as they apply to rhinoceros and examines the evidence that rhinoceros are more prone to extinction than are smaller species of herbivorous mammals.

Seven basic questions must be answered to save populations of large endangered species from extinction: What is their general demographic profile and the trajectory of the population? How prominent is human-induced mortality, particularly in adult animals? What is the level of genetic variability in the population, and has a population bottleneck affected it? Is the species a good or a poor disperser? Do populations occur at low densities, thereby requiring large areas to maintain viable populations? Is the species highly sensitive to human disturbance, or can it forage and survive in less pristine or even degraded landscapes? And when populations of endangered large mammals decline dramatically, at what point do they become "ecologically extinct"? In other words, is there a population threshold below which endangered species may still persist but no longer play their normal ecological role — as consumers of plants, dispersers of seeds, or regulators of other species populations — in the ecosystem to which they belong?

CHAPTER 4

SIZE AND SEXUAL DIMORPHISM IN GREATER
ONE-HORNED RHINOCEROS

Prior to late-Pleistocene and Holocene megafaunal overkill, nearly every ecosystem on earth included one or more very large herbivores that were too big (at least as adults) to be killed by the largest carnivores in the system. In Africa, there are rhinos and hippos, in addition to elephants, that enjoy immunity from lions. In the north, adult moose repel wolves, and in the Neotropical forest, tapirs shrug off jaguars. Even on some remote islands, Madagascar had its elephant birds, New Zealand its moas, the Antilles its ground sloths, and the Seychelles, Galápagos, Aldabra, and New Caledonia their tortoises.
—John Terborgh, *Requiem for Nature*

WE DARTED OUR FIRST GREATER one-horned rhino, a large male, in the winter of 1985. Within a few minutes, my col-.leagues and I saw him ease to the ground as the powerful morphine derivative worked its magic. We descended from our elephants to measure this magnificent animal and to attach a radio collar. Like Lilliputians next to Gulliver, we stared in wonder at the behemoth that lay before us (figure 4.1). Our hearts stopped suddenly as the supposedly sedated rhino stood up and rumbled off into the forest. Our textbook on animal capture omitted mention of this behavior. As the year progressed, we became more experienced at immobilizing rhinoceros. With great relief, we found that our first alarming encounter was an anomaly. But at the moment that the very first rhinoceros jumped to its feet and scared the daylights out of us, I had the uncomfortable and rather unscientific thought that our powerful modern drugs were somehow ineffective for sedating prehistoric large mammals. And, most of all, it is startling to discover just how big greater one-horned rhinos are (figure 4.2).

I was especially intrigued by two questions concerning the size

FIGURE 4.1 Researchers begin to measure a sedated greater one-horned rhinoceros (the author is in the foreground). (Courtesy John Lehmkuhl)

FIGURE 4.2 An elephant driver crouches next to the head of a large female rhinoceros, giving some idea of just how large an animal the rhinoceros is. (Photo by Eric Dinerstein)

of greater one-horned rhinos. Among mammals, the males of most species are larger than the females. The divergence reaches a maximum in a few species, such as elephant seals, where males can be more than twice the size of females. Large body size may be under strong selection pressure, resulting in only the biggest males gaining access to breeding females. Early on, I also wondered whether only the largest male rhinoceros were responsible for all the breeding in the population. Large body size may also be selected for as a way to discourage predators. But how big does an ungulate on Chitwan's floodplain have to be to, in the words of John Terborgh, be "too big (at least as adults)" to be killed by tigers, the largest carnivores in the ecosystem?

Male ungulates can also differ in appearance from females in other conspicuous ways. Such features as presence and size of horns or antlers, elongation of canine teeth, and coloration may be under natural selection and closely linked to reproductive success (Clutton-Brock 1988). Other members of the odd-toed ungulates (tapirs and horses), however, show little dimorphism in such traits (Berger 1986). How and why do male and female rhinos differ in terms of size, horns, and other features?

To address these questions I measured thirty-eight body features of fifty free-ranging rhinoceros.* Data from wild populations are essential because the superior diets available to captive-born individuals are likely to influence measurements recorded from captive populations. Measuring representatives of all sex and age groups allowed me to classify animals into discrete age classes based on body features and dental definition.

Close-up examination of dentition, obtained during immobilization, was important for correlating dental wear with other signs of age in adults (e.g., facial wrinkles, erosion and width of the base of the horn, accumulation of scars). Once we figured out the correlation, we were able to age individuals by observing these features through binoculars and did not have to immobilize them to do so. One implication of this technique is that because adults exhibit a high level of individuality in a variety of body features, it is possible to identify individuals with a high degree of accuracy. As in counting elephants, giraffes, whales, and other large mammals, the ability to identify individuals makes counting rhinoceros much easier, allowing us to track their numbers and potential for recovery.

---

*For the methods used to capture greater one-horned rhinoceros, see Dinerstein, Shrestha, and Mishra (1990).

How Big Are Greater One-Horned Rhinoceros?

The best way to appreciate the size of greater one-horned rhinoceros is in relation to that of other megaherbivores. White rhinoceros are the largest of the five extant species, and they are the third-largest land mammal (after African and Asian elephants); field weights are 2,000 to 2,300 kg for adult male rhinos from Natal and about 1,600 kg for females (Owen-Smith 1988). Greater one-horned rhinoceros rank as the second-largest species of rhino and the fourth-largest living terrestrial mammal. We captured a large adult male for translocation to Royal Bardia National Park in 1986 that exceeded 2,000 kg, indicating some size overlap with the white rhinoceros. Based on measurements of other males, I suspect that this is close to an upper limit in body mass. In captivity, adult male greater one-horned rhinoceros may reach 2,100 kg. The biggest black rhinoceros reach 1,300 kg. Javan rhinoceros are reported to be about the same size as black rhinoceros (Owen-Smith 1988), but no published account of wild-caught animals from Java or Vietnam is available. The Javan rhinoceros in Vietnam is thought to be smaller than the rhino native to Indonesia, based on photographs and the size of footprints. Sumatran rhinoceros, at about 800 kg, are tiny in comparison.

Body Size in Relation to Age Categories

Rhinos vary considerably in size in accordance with their age. Not surprisingly, mothers dwarf their newborn calves. Our study (reported here) considered animals to be calves until they separated from their mothers, at about the age of four. After four years of study and continued monitoring of registered females, we confirmed this classification. Subadults are at least four years old and younger than six (figure 4.3). At the upper end of this age class, subadult females were occasionally difficult to differentiate from young adult females that were small in total body length. Subadult males began to show development of secondary neck folds. Behavioral characteristics also allowed Laurie (1978) and me to identify subadults. Subadult males and subadult females frequently form same-sex groups. Males probably do so for increased ability to detect, and be protected from, dominant males. Finally, subadults typically try to remain close to their mother even after a sibling is born. This behavior provides an opportunity to assess the size of a four- to five-year-old calf against that of the mother (figure 4.4).

A

B

FIGURE 4.3 Subadult (A) male and (B) female rhinoceros. They are at least four years old but younger than six. (Photos by Eric Dinerstein)

A

B

FIGURE 4.4 Females, with calves in four age categories based on calves of known birth dates: (A) mother with a calf younger than one year old; (B) calf aged one to two years trotting ahead of its mother; (C) calf aged two to three years at left; (D) calf aged three to four years at left. (Photos by Eric Dinerstein)

C

D

Other features helped me to estimate the age of adults. I constructed three age categories for adults based largely on dental condition and horn length and then related these to other characteristics (box 4.1). Although quite complex, these features are essential for accurate assessments of demographic characteristics of the population.

Young adult males and females show considerable overlap in measurements. Because my samples were too small to allow me to detect differences, I combined the sexes for this category. The combined data show that young adults are noticeably smaller than intermediate-age and old adults, indicating that females are reproductively active while still growing (appendix B, table B.1).

BOX 4.1 *AGE CATEGORIES FOR ADULTS*

From a distance, young adults (roughly, 6–12 years) appear to be free of physical defects—they exhibit few scars or wrinkles. On closer inspection, they show little or no wear of the molar teeth, the essential machinery for breaking up the massive amounts of grass tissues that these animals ingest. Thus young adults possess sharp occlusal ridges on lower molars, and their lower outer incisors were less than 5 cm long and less than 3 cm wide at the base (males only). The horn is intact, with no erosion around the base, and short (15 cm for females; 18 cm for males). Young adults develop secondary shoulder and neck folds that are relatively small (figure B4.1a).

A

B

FIGURE B4.1 Young adult rhinoceros. (A) Note the beginning of secondary neck skin folds in the male. (B) A young adult female with her calf. (Photos by Eric Dinerstein)

A

B

FIGURE B4.2 Intermediate-age (A) male and (B) female rhinoceros. (Photos by Eric Dinerstein)

BOX CONTINUED

Intermediate-age adults (figure B4.2), which are older than twelve and younger than twenty-four, show moderate wear on the occlusal surface of the lower molars with mandibular outer incisors that are 4 to 5 cm long and more than 3 cm wide at base (females) and more elongated in males (range, 4.5–8.7 cm). Horns show moderate erosion at the base and are 20 to 28 cm long in females and 20 to 30 cm long in males and mostly intact (i.e., no broken tip). Circumference at the base of the horn is 48 to 54 cm for males. Intermediate-age adults are longer and of greater girth than most young adults. Males show extensive development of neck, shoulder, and secondary shoulder skin folds and thick upper-neck muscles. Both sexes have moderate wrinkles around the mouth, under the zygoma, and around the eyes and forehead. Deep scars on anal folds, face, and back of legs are uncommon. A female that has successfully weaned its first calf and is nursing its second is considered an intermediate-age animal.

The wear of courtship and combat and the effect of years of grinding silica-laden grasses between their molars help distinguish old adults (older than 25 years) from rhinoceros in other age categories. Old adults have heavy wear on the molars with well-formed depressions on the occlusal surface. Lower outer incisors are 4 to 5 cm long and more than 3 cm at base (females) and elongated in males (range, if entire, 5.1–8.9 cm) (figure B4.3) or often broken. The horn is either long (20–33 cm) or broken and heavily worn and eroded, often with deep anterior and occasionally with posterior grooves. In males, horn base is large in

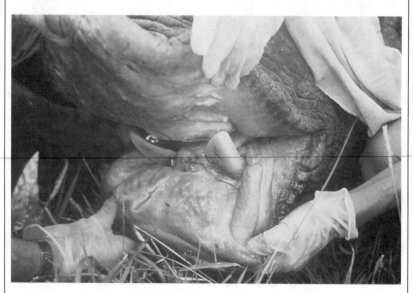

FIGURE B4.3 The procumbent lower outer incisors (tusks) are the prime weapons that males use in intraspecific combat. The individual pictured here had some of the longest incisors that we measured. (Photo by Eric Dinerstein)

A

B

FIGURE B4.4 Old adult rhinoceros. (A) The male is larger around the neck and shoulders than the female. (B) Abire Pothi, an old female, was breeding in 1975 (Laurie registered her in 1978) and was still bearing calves in 1998. She is easily identified by a pink forefoot. (Photos by Eric Dinerstein)

BOX CONTINUED

circumference (55–95 cm), commonly with signs of breakage and subsequent regrowth. Old males develop extensive neck, shoulder, and secondary shoulder skin folds and thick upper-neck muscles (figure B4.4). Old adults have deep wrinkles around the mouth, under the zygoma, and around the eyes and forehead. Scars are prominent around the anal folds and backs of legs, and often the second cross-skin fold bears major scars, while minor scars appear on the face and other areas. Shoulder girdles, hip girdles, and ribs are prominent in old females. An old adult often has a cut ear or pieces missing in one ear and occasionally both. Unlike young and intermediate-age males, old males seldom run from human observers riding on domesticated elephants, and most old males squirt urine back between their legs in a dominance display (figure B4.5).

FIGURE B4.5 A dominant male engages in squirt-urination. This behavior involves inverting the penis and propelling urine in a stream between the hind legs for several meters. (Photo by Eric Dinerstein)

Variation in Measurements and Physical Characteristics

I found no significant differences between intermediate-age and old males (n = 8 and 8, respectively) or between intermediate-age and old females (n = 4 and 5, respectively) for any of the twenty-three measurable characteristics (Dinerstein 1991a). To increase the number of samples that I had for comparisons between the sexes, I combined data within each sex and renamed the combined category as "older adults." For males, this new classification parallels breeding activity; males classified as young adults did not breed in the study area between 1984 and 1988.

Among older adults, males are significantly larger than females in neck circumference just behind the head, neck circumference at the shoulder, maximum skull circumference, and neck width behind the head (see the statistical notes at the end of this chapter). Differences approach significance levels for cranial breadth and for neck width at the shoulder. Larger neck measurements in breeding males are attributed to the greater muscle mass and enlarged skin folds around the neck (bibs). Breeding males require substantially larger radio collars than do females. Larger samples probably would show that older males are taller than females at the shoulder, although chest circumference may prove to be more variable. Older males are slightly longer in length of head and body than females, but this was not significant.

We observed broken incisors in four of the eight older males. If we remove individuals with broken incisors from the sample of older males, males have significantly longer lower outer incisors than females, as measured by the longer of the two incisors or by the mean length of the pair of intact incisors. In contrast, lower outer incisors of young adult females are as long as or longer than incisors of young adult males. Horn circumference at the base also is significantly larger in males. Length of horn is roughly equal (appendix B, table B.1). Subadults (4–≥ 6 years) are readily distinguished from young adults by smaller size and shorter length of horn.

Variation in easily recognizable body characters — such as epidermal knobs, extra skin folds, horn and ear shape, and scarring — permits rapid identification of individuals (Laurie 1978) (figure 4.5). Variation in some features — such as cuts, torn ears, and other signs of combat among adult males — helps us understand the relationship among body size, age, and dominance. In the field, for example, we can readily determine which males are old and which are intermediate age. Old males are far more likely than intermediate-age males to have cut, scalloped, or torn ears; broken horns; scars on the second cross-skin fold; and erosion of the horn at the base (appendix B, table B.2). We found no intermediate-age males with broken horns. All the oldest males show some scars on their flanks, around the tail, or on the rear-leg folds, results of intense fighting among dominant breeding males. They also have pronounced secondary shoulder and neck folds.

A

B

FIGURE 4.5 Variations in and combinations of easily recognizable nonmensural characteristics permit the rapid identification of most individuals from a photo-identification register (after practice) in the field. These characteristics include (A) the size and location of epidermal knobs (also note the crooked tail in this female); (B) extra skin folds on the neck; (C) the integrity of the horn (entire, grooved, eroded at base, broken off); and (D) the integrity of the ears. (Photos by Eric Dinerstein)

C

D

Old females are less likely to have cut, scalloped, or torn ears than males. Old females are more likely to have heavy erosion of tissue around the horns than intermediate-age females, although this difference only approaches significance. However, erosion around the horn is one of several traits (e.g., extensive old scars on the rear, heavy wrinkles on the face, length of horn, known age of previous calves) that help to place females in the old or intermediate-age classes.

Nearly half of old adults develop a longitudinal groove in the horn. Almost half of old females develop epidermal knobs (3–25 cm long) on the posterior cross-skin folds. Epidermal knobs are significantly less common on young adults than on older adults, suggesting that they may appear as the animal ages. The proportion of adults in each age class with epidermal knobs on anal folds was about equal.

Sexual Dimorphism in Rhinoceros

Horns Versus Lower Outer Incisors

All three species of rhinoceros in Asia have small horns in comparison with the two African species. The two African species fight and display exclusively with the horns and lack lower outer incisors. The Asian species, in contrast, possess elongated lower outer incisors, although they are smaller in Javan and Sumatran rhinoceros than in the greater one-horned rhino. These dental weapons are the most conspicuous differences between male and female greater one-horned rhinos. They are used frequently in intermale fights that determine dominance and access to estrous females. In contrast, female greater one-horned rhinos often have longer horns (although slightly narrower) than do the males that breed them. Three dominant males during my study in Chitwan maintained their status with broken horns but intact incisors.

Aspects of breeding biology in greater one-horned rhinoceros provide a background for understanding patterns of sexual dimorphism as they relate to incisor size. Male greater one-horned rhinoceros form dominance hierarchies, and an individual's tenure as alpha male is short in comparison with its lifespan (chapter 7). Limited chances for copulation probably heighten aggressive behavior when a female cycles into estrous. Under this regime, selection may be strong for long outer incisors (or tusks) to gain access to areas where these females congregate. Androgen levels in greater one-horned rhinoceros may control the elongation of the incisors in much the same way that some primates show dimorphism in canine length (Zingeser and Phoenix 1978). Courtship in greater one-horned rhinoceros is among the most violent for mammals; males aggressively pursue females during long courtship chases (>2 km) and attack females with their incisors or ram into them in an attempt to subdue them.

Body Mass and Other Features

Larger samples probably will show that old adult male greater one-horned rhinos are taller at the shoulder and slightly heavier than most old females. However, slight differences in body mass may not be a critical dimorphic feature in species whose adults exceed 1,500 kg. Three of the nine older females measured in this study were longer, and they appeared to be almost as large as the males that bred them.

Among the other rhinoceros, free-ranging black rhinoceros show no difference in body mass between the sexes (Freeman and King 1969), male white rhinoceros are estimated to be about 28 to 41% heavier than females (Owen-Smith 1988), and wild-caught adult Sumatran rhinoceros show no obvious size dimorphism (Thomas Foose and Michael Dee, personal communications, 1987). No data exist for Javan rhinoceros.

In captivity, male greater one-horned rhinoceros can be 1,000 kg heavier and at least 25 cm taller than females (Lang 1961; Dee, personal communication, 1987). However, I have never observed this extreme size dimorphism in free-ranging animals. Body-mass data for male zoo animals must be viewed with caution because zoo-born animals probably have a significantly more nutritious diet than their wild counterparts. Males born in captivity may become much larger than captive-born adult females after only four years. In the wild, four-year-old males are always substantially smaller than adult females.

Dimorphism also is observed in the massive neck and upper shoulder muscles, which are more extensively developed in adult males. These muscles provide the force behind the slashing and gouging with the incisors. The extensive primary and secondary neck and shoulder folds found in dominant males may serve for display in head-to-head confrontations between rival males. The greater one-horned rhinoceros is believed to have poor eyesight, but the head-on display, which often precedes combat, occurs when males are within a few meters of each other. Presenting the large neck and shoulder skin folds and baring the outer incisors during display may cause a challenger to submit and flee before the encounter with the dominant male intensifies into a fight. These neck folds may also deflect the penetration of an opponent's incisors from the neck, chest, and shoulder area during actual confrontation. This is the region where most severe attacks first occur before one male inevitably turns and runs from the other.

Reduced size dimorphism of wild greater one-horned rhinoceros probably results from forces operating on both sexes. Genetic factors presumably determine the upper limit of body mass, but comparing observations of wild and captive individuals implicates diet and behavior in influencing the degree of dimorphism in this feature. There may be selection for large females, as larger females would be better able to defend themselves and their offspring against males, a special case of the "big mother" hypothesis (Ralls 1976). More likely, poor nutrition and behavioral factors may restrict growth of wild males. Data for several mammalian species show that size of adult males is reduced more severely than size of adult females if food resources are curtailed during the normal period of rapid growth (Ralls and Harvey 1985; Widdowson 1976; Wolanski 1979). Stunted growth for some males may be related to reduced protein intake during the long period of adolescence. Males occupy marginal habitats with poor-quality forage during the nearly ten-year period between their physiological capability to breed (about age 5) and their actual association with females on prime grazing areas (about 12–15 years). Stress associated with frequent harassment and attacks by breeding males on younger males may also result in free-ranging males' failing to reach their potential size, whereas they will in captivity because males are held separately.

Greater one-horned rhinoceros rank among the largest of all land mammals, surpassed by only the two elephant species and the white rhinoceros of Africa. Among the ungulates of Chitwan's floodplain, rhinoceros, which are cecal digesters, dwarf the smaller ruminants in body mass. They are also far more abundant than Indian bison (another ruminant) and Asian elephants (another hind-gut fermenter). However, unlike the elephants, white rhinoceros, and Indian bison, greater one-horned rhino males are not significantly larger than females in body size. Despite the mythology surrounding its single horn, it is the less conspicuous, but far more lethal, procumbent lower outer incisors that are the most important physical feature for which males and females show significant dimorphism. The size and condition of incisors affect many aspects of their ecology, and I explore this aspect in subsequent chapters.

Statistical Notes

Variation in Measurements of Adults Relative to Age and Sex

Among older adults, males were significantly larger than females

1. Neck circumference behind the head (Mann-Whitney $U = 112$, $N = 9$ and 13, $p < 0.001$).
2. Neck circumference in front of the shoulder ($U = 106$, $N = 9$ and 1, $p = 0.001$).
3. Maximum skull circumference ($U = 79.5$, $N = 9$ and 11, $p < 0.005$).
4. Width behind the head ($U = 63.5$, $N = 7$ and 11, $p < 0.05$).
5. Differences approached significance levels for cranial breadth ($p = 0.10$).
6. Differences approached significance levels for shoulder height ($U = 109$, $N = 4$ and 3, $p = 0.10$).

For mandibular outer incisors (tusks) and horns, males have

1. Significantly longer mandibular outer incisors than females as measured by the longer of the two incisors ($U = 109$, $N = 9$ and 13, $p < 0.001$).
2. Significantly longer mandibular outer incisors than females as measured by the mean length of the pair of intact incisors ($U = 112$, $N = 9$ and 13, $p < 0.001$).
3. Significantly larger horn circumference at base than females ($U = 93.5$, $N = 9$ and 11, $p < 0.001$).

Variation in Physical Characteristics

1. Old males were far more likely to have cut, scalloped, or torn ears than intermediate-aged males ($X^2 = 4.3054$, d.f. $= 1$, $p < 0.05$).

2. Old females were less likely to have cut, scalloped, or torn ears than males ($X^2 = 8.5479$, d.f. $= 1, p < 0.005$).

3. Old females were more likely to have heavy erosion of tissue around the horns than intermediate-age females, although this difference only approached significance ($X^2 = 2.2432$, d.f. $= 1, p < 0.10$).

4. Epidermal knobs were significantly less common on young adults than on older adults ($X^2 = 7.2465$, d.f. $= 1, p < 0.01$).

CHAPTER 5

THE BIOLOGY OF AN EXTINCTION-PRONE SPECIES: FACING DEMOGRAPHIC, GENETIC, AND ENVIRONMENTAL THREATS

Extinction is impartial in choosing its victims; species of all sizes, trophic levels and taxonomic groups fall prey. Rarity proves to be the best index of vulnerability. Rare species include top predators, and frequently other large species as well as habitat specialists.
—John Terborgh and Blair Winter, "Some Causes of Extinction"

ON A HOT SPRING AFTERNOON in late March, I was riding elephant-back to a grassland known as Pipariya in the eastern part of Chitwan. With me were three Malaysian biologists who had devoted six years to studying and capturing the elusive Sumatran rhinoceros. Despite extensive tracking, they had yet to see a single Sumatran rhino in the wild other than those that had been captured in pit traps. As we emerged from the forest, I pointed out the gray, boulderlike objects dotting the grassland. Fires had burned through two weeks earlier, and the abundance of new shoots had attracted a congregation of greater one-horned rhinos. Mothers and calves seemed to be everywhere. Within the span of two hours, in an area no bigger than a shopping-mall parking lot, we counted thirty-five individuals, the highest concentration I had ever seen. We got close enough to one individual that my colleagues could take photographs of the body folds, horn, and rivetlike tubercles on the hind legs. After posing for a few minutes, the rhino was startled by an elephant and trundled off, uttering deep huffing grunts in cadence with its departing trot.

Sometimes, because I have observed rhinoceros on thousands of occasions, and because they are locally abundant, I forget just how rare globally these creatures truly are. I often need the awe of scientists and other visitors to help me regain my perspective. The odd-looking appearance of greater one-horned rhinoceros, accentuated by the armorlike skin folds, prehensile lips, and nasal horns, is quite fantastic. An ancestral rhino, *Elasmotherium*, with a single large horn perched on its forehead, showed some resemblance to the mythical unicorn (see figure 1.1). It likely went extinct in the late Pleistocene, a period when a variety of large mammals disappeared.

The accelerating decline of most megaherbivore populations today raises several profound questions: Are very large mammals anachronisms in human-dominated landscapes? Do they belong to Terborgh and Winter's (1980) list of extinction-prone species? I have suggested that their large body size and high mobility contributed to the dominance of rhinoceros during much of the age of mammals. Now that remaining species must survive in fragmented habitats, are certain life histories and behavioral features associated with very large body size no longer an asset for survival? If today's rhinoceros lacked horns, like some of their ancestors, or if populations could be dehorned to reduce their commercial value, would they still be vulnerable to extinction?

Demographic Threats of Extinction

Scientists have widely accepted the notion that small populations, typically fewer than 100 individuals, face a serious threat of extinction, particularly if populations remain at low levels for extended periods. When populations are at extremely low levels, only a few females may be reproductively active at a given time. In some years, these females might give birth to all males or all females, or under certain circumstances they may not reproduce at all. This last situation is particularly prevalent if an endangered population is widely scattered and potential mates have difficulty finding each other when females are receptive. For many large mammal populations today, human-induced mortality from poaching or habitat loss confounds interpretation of the demographic effects of small population size. Limited empirical data on free-ranging populations of large mammals exist to test the theory that small populations are in greater demographic jeopardy than are large populations. A study of 122 mountain sheep (*Ovis canadensis*) populations in five U.S. states found that all populations numbering fewer than

50 individuals went extinct within fifty years, regardless of habitat (Berger 1990). Populations larger than 100 individuals showed no predilection for extinction. Some have challenged Berger's (1999) extinction model (Wehausen 1999), but it remains as a useful precautionary threshold for conservation biologists and wildlife managers.

Most rhino populations in protected areas around the world do not exceed the threshold of 100 individuals. For the three Asian species, most populations fall below Berger's danger zone of 50 individuals (Foose and van Strien 1997). Consequently, the threat of persistent small population size has been a central theme in various editions of the Asian Rhino Action Plans prepared by the International Union for Conservation of Nature and Natural Resources (Foose and van Strien 1997).

The Trajectory of Chitwan's Rhinoceros Population During the Twentieth Century

In the early twentieth century, the estimated number of rhinoceros in the Chitwan Valley exceeded 1,000 individuals. Populations appear to have remained at this level until the early 1950s, when the eradication of malaria opened the floodgate to human settlement. By 1966, the total rhino population was believed to have declined to 60 to 80 individuals (figure 5.1) because of land clearing and heavy poaching. Laurie's census in 1975 reflected a dramatic turnaround, yielding a total population estimate of 270 to 310 individuals (Laurie 1978). Between 1984 and 1988, I repeated Laurie's census using similar techniques but with more staff and more elephants to

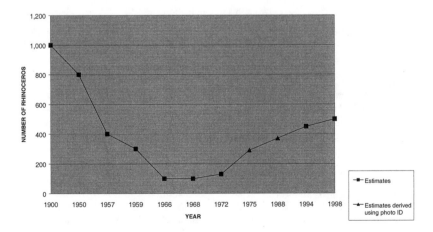

FIGURE 5.1   The Chitwan rhinoceros population during the twentieth century.

search for rhinoceros (appendix A). The dense vegetation of Chitwan, logistics, and the large size of the park precluded censusing each section with equal intensity. Moreover, natural breaks in preferred rhino habitat separate animals into what can be treated as subpopulations, ranging from the Kagendramali subpopulation in the east, through the Sauraha subpopulation in the eastern and central parts, the Bandarjhola Island and Narayani River subpopulations, to the West subpopulation (table 5.1; figure 5.2). Both Laurie and I used these delineations to census subpopulations. I derived a minimum estimate of 358 individuals and a maximum of 376. A count conducted in 1994 estimated 446 to 466 animals using even more elephants and searchers than in either Laurie's or my study but over a much shorter search time (Yonzon 1994). An attempt at a total count in the spring of 2000 placed the population at more than 500 for the first time since the 1950s. The recovery of greater one-horned rhinos demonstrates that populations can rebound vigorously from heavy poaching when provided with sufficient habitat and strict protection.

The Sauraha subpopulation increased by eighty-six animals (49%) in the thirteen years from 1975 to 1988. The growth rate calculated from regression analysis for the Sauraha subpopulation (1985–1988) indicates an

TABLE 5.1.   Estimated Total Population of Greater One-Horned Rhinoceros in Royal Chitwan National Park, April 1988

Area	Number
Sauraha arid Kagendramali subpopulation	252
Relocated animals (1986–1988) (from this subpopulation only)	−24
Subtotal	228
West subpopulation (1986)	61
Animals assumed to have been missed (7%) during 1986	4
Estimated population increase, 1986–1988	7
Adjusted subtotal, West population	72
Bandarjhola Island and Narayani River subpopulation	34
Animals assumed to have been missed (7%) during 1986	2
Estimated population increase, 1986–1988	4
Adjusted subtotal, Bandarjhola Island and Narayani River	40
Outlying areas (Ramoli, Tikoli, Botesimra)	18
Total minimum estimate for 1988	358

NOTE: Population excludes thirty-four animals translocated between 1980 and 1988.

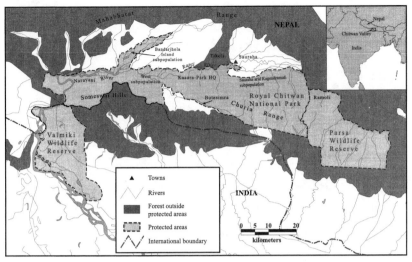

A

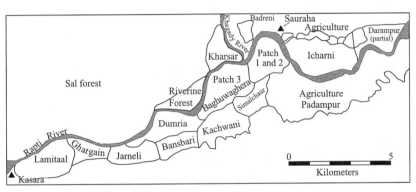

B

FIGURE 5.2 (A) Major features and locations of the four rhinoceros subpopulations—Sauraha, West, Bandarjhola-Narayani, and south Botesimra—in and near Royal Chitwan National Park. (B) The detailed map of the Sauraha area shows the blocks in which we searched for greater one-horned rhinoceros.

annual increase of 4.8% (figure 5.3). The population growth rate estimated from schedules of fecundity and survivorship for the same period indicates an annual increase of 2.7% (appendix C, table C.1). When I combined the 1975 data from Laurie (1978) with my data and did a regression analysis, I found that the growth rate of the Sauraha subpopulation from 1975 to 1988 was 2.5% ($r^2 = 0.961$, $p < 0.01$). (For more details on the use of life tables for calculation of r, see appendix A.) In contrast to Sauraha, the West subpop-

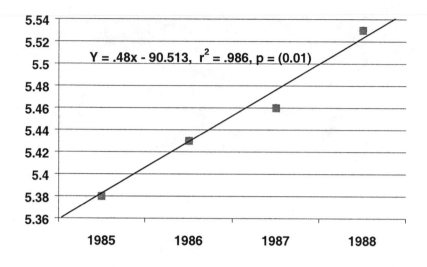

FIGURE 5.3 Population growth rate of greater one-horned rhinoceros in the Sauraha subpopulation at Royal Chitwan National Park, 1984–1988.

ulation increased by only 2.2% between 1975 and 1988, for a mean annual rate of increase of 1.7%. Comparisons were not possible for the Bandarjhola and Narayani subpopulation because population size was estimated by photo registration in the 1988 study, whereas Laurie used dung piles and footprints.

Laurie and I each spent four years counting rhinos using a photo-registration technique, whereas Yonzon (1994) used a modification of this technique for a fifty-day period. Fortunately, Yonzon's team counted rhinoceros during the period of the year when census by elephant is most feasible (the hot-dry period), and an abundance of new grass shoots drew rhinoceros to grasslands in the afternoons, when they are most visible and least likely to run. However, Yonzon found that a large number of individuals either had no distinguishable markings (25%) or were difficult to recognize as they rushed by (11%). This suggests that Yonzon may have counted some fraction of animals more than once. Both Laurie and I found that subadults are difficult to positively identify because they often lack distinguishing features. My study found the highest proportion of subadults in the Bandarjhola block; this is where Yonzon's count and the 1988 census diverge the most. In the West and Bandarjhola–Narayani subpopulations, rhinoceros tend to flee from approaching elephants more readily than in the Sauraha subpopulation, so the potential for a sampling bias to develop is great. The inability to easily distinguish subadults, together with sampling bias, may have significantly inflated estimates in the 1994 count.

In 1988 it seemed to me that Chitwan could support at least another 100 individuals, to a population size exceeding 500. Several large tracts of *Saccharum spontaneum* grasslands suitable for maintaining high densities of rhinoceros were underpopulated. Harassment by cattle herders may have kept rhinoceros from occupying these areas, but, beginning in 1988, these areas were zoned strictly for wildlife. The latest tally from Chitwan in 2000 realized this potential. Now the opportunity exists to restore rhinoceros habitat in Chitwan's 750-km² buffer zone, which would facilitate population expansion in the coming decades. Based on observations in existing restored buffer-zone forests, the Chitwan population could reach 800 individuals if local residents can restore approximately two-thirds of the buffer zone (for more data, see chapter 10).

Recovery of Other Rhinoceros Populations

Data on recovery of other rhino populations in large reserves comes from South Africa (Owen-Smith 1981). The Umfolozi white rhinoceros population grew at a constant rate of 9.6% per year from 1969 to 1972. High birthrates and a short interbirth interval ($\overline{X} = 2.2$ years) account for this high rate of increase. The interbirth interval for the Umfolozi population is half that of the Chitwan population, even though the white rhinoceros is slightly larger in body mass.

Black rhinoceros populations kept in well-guarded game ranches have shown remarkable increases over short periods, in some cases exceeding 13% per year. The Solio Ranch in Kenya has experienced population growth of 4 to 10% per year, and in small reserves such as Nairobi National Park growth has been as high as 10% per year (Brett 1998). These small populations quickly reach carrying capacity, requiring additional translocations. We lack data that would determine long-term trends for free-ranging black rhino populations in larger reserves.

Demographic Features of the Sauraha Subpopulation

Longitudinal studies of demography help us understand how quickly or slowly endangered populations of large mammals recover from episodes of near extinction. Because most of the Chitwan population is concentrated in the Sauraha subpopulation, both Laurie and I concentrated much of our census efforts there. Consequently, we have the most reliable data from the Sauraha area for comparing demographic features from 1975 and 1988. Laurie (1982) estimated that the Sauraha subpopulation in 1975 contained 176 animals. In 1988 the Sauraha subpopulation contained 228 individuals, 60 to 63.5% of the total estimated population.

Sex and Age Structure and Breeding Status

Between 1984 and 1988, in the Sauraha subpopulation, sex ratios for calves, subadults, young adults, and old adults were not significantly different from parity ($p > 0.05$) (table 5.2). Intermediate-age adult females (12–20 years) were significantly more numerous than males (see the statistical notes at the end of this chapter). However, adults in the West and Bandarjhola–

TABLE 5.2.   SEX AND AGE COMPOSITION OF GREATER ONE-HORNED RHINOCEROS IN THE SAURAHA SUBPOPULATION OF ROYAL CHITWAN NATIONAL PARK, 1975 AND 1988

Parameter	1975	1988
Adult sex ratio (% M)	34.1	39.9
Subadult sex ratio (% M)	55.3	54.2
Subadult and adult combined sex ratio (% M)	50.9	63.5
Subadult as % of N	22.8	13.3
Adult F as % of N	33.5	38.2
Adult F with calves as % of N	26.3	20.9
Adult M as % of N	17.3	25.3
% of adult M known or assumed to have bred during study period	—	48.3
% of adult F with calves	78.6	59.8
% of adult F with calves, excluding F 6–7 yr old	—	77.0
% of adult F (6–12 yr) with calves	—	54.1
% of adult F (7–12 yr) with calves	—	90.9
% of adult F (12–20 yr) with calves	—	64.3
% of adult F (>20 yr) with calves	—	63.6
% of N <12 yr old	—	62.0

NOTE: The 1975 data come from Laurie (1978). Laurie classified subadults as 3–9 years old, whereas we classified subadults as 4–6 years old. Total population (N) was 275 for 1975 and 375 for 1988. M, male; F, female.

Narayani subpopulations showed no significant differences in sex ratio (appendix C, table C.2). Subadults accounted for only 8% of the registered West subpopulation but 30% of the Bandarjhola and Narayani subpopulations and 13% of the Sauraha subpopulation. There appears to have been a higher percentage of females with calves in 1975 than in 1988. However, in Sauraha, if adult females younger than six or seven are excluded in calculating the 1988 data (table 5.2), the proportion of females with calves is the same. Results from 1988 also show that Sauraha's percentage of subadults decreased.

Younger, stronger males frequently attacked old males (older than 20 years) that remained near high concentrations of breeding females (Laurie 1978). In at least five instances during my study, such attacks proved fatal. Another five breeding males suffered serious wounds in fights and retreated to areas with low densities of breeding-age females, where the most aggressive males did not venture. I estimated that 48% of adult males in the Sauraha subpopulation ($n = 28$) mated during the study period (chapter 7).

Fecundity and Mortality Rates

SEASONALITY OF BIRTHS. Temperate-zone ungulate species demonstrate strong seasonality in reproductive schedules and parturition (period of birth). Patterns in tropical ungulates are less distinct but also less well studied. Greater one-horned rhinos live in subtropical habitats marked by a long dry period, so one might suspect strong seasonality in reproduction. One might also suspect that selection would favor avoiding parturition during the late monsoon season because severe floods could drown calves or separate them from their mothers, leaving calves vulnerable to predation by tigers. Fifty-three calves were born in the Sauraha subpopulation during the study period (figure 5.4). However, I could detect no significant temporal pattern in the distribution of births throughout the calendar year. To increase my sample size, I reclassified monthly birth data from my study period in the same bimonthly format used by Laurie (1978) for 1972 to 1975, and the combined data set ($N = 113$) also did not reveal any significant seasonality. One advantage of aseasonality in the recovery of an endangered population is that breeding can occur in any month if calf mortality occurs.

AGE AT FIRST REPRODUCTION. The higher the age at first reproduction, the longer it takes for a population to rebound. In this regard, rhinoceroses

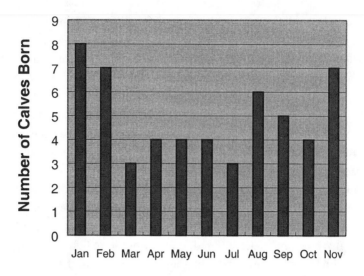

FIGURE 5.4 Births of greater one-horned rhinoceros by month in the Sauraha subpopulation at Royal Chitwan National Park, 1985–1988.

are truly at a reproductive disadvantage compared with smaller mammals, most of which breed after the first year and in some instances give birth before 1 year of age. I estimate that the mean age at first birth for two known-age females in the Sauraha population was 7 to 7.5 years. An additional fifteen adult females (40%) had not given birth to their first calf by the end of my study. Of these, I classified thirteen as 6 to 7 years old, one as 7 to 8 years old, and one as younger than 8 years old. However, females are long lived and seem to produce calves throughout their lives, so longevity may at least partially compensate for a slow age until first reproduction. Several females in Chitwan that were part of Laurie's adult breeding population in 1972 were also part of my "old adult" breeding population sixteen years later, and they were still breeding as of 1998 (see figure B4.4, bottom).

INTERBIRTH INTERVAL. Another demographic variable affecting recovery is the interbirth, or intercalving, interval. Assuming that food resources are abundant and populations below carrying capacity, one might expect rhinoceros to produce calves at a rate of one every 2.5 years. This seems even more plausible, given the short intervals exhibited by the larger white rhinoceros of Africa, which lives in a more xeric habitat.

My study extended long enough to record two births each for sixteen of eighty-seven adult females. The interbirth interval based on thirteen animals whose calves survived to independence was 45.6 months ± SD 1.8 months (range = 34–51 months).* Older females tended to exhibit a higher interbirth interval. Six females had interbirth intervals longer than 48 months; one was twelve to twenty years old, and five were older than twenty. Three of the sixteen females exhibited short interbirth intervals, giving birth again at 17, 22, and 31 months. However, each had an extenuating circumstance surrounding the subsequent birth: the loss of a calf probably less than a few months old. A dominant bull trampled the first calf of one female, another calf was captured for shipment to a zoo, and the third calf was thought to have died or been separated from its mother. Apparently, the death of a calf can shorten the interbirth interval.

An additional twelve females exhibited much longer interbirth intervals ($\bar{X}$ = 60.9 ± 3.4 months; range = 48–88 months). They were accompanied by calves of an advanced age and did not give birth during the 48 months of study. Six females were categorized as twelve- to twenty-year-olds and six as older than twenty. Some of the oldest females may no longer have been reproductively active. Some may have aborted or given birth to a calf that died before we registered it, but this is unlikely. The mean interbirth interval for rhinoceros in 1988 in Chitwan exceeded 48 months, in contrast to Laurie's estimate of 42 months during 1975.

AGE-SPECIFIC BIRTHRATE. Intermediate-age adult females (12–20 years) had higher birthrates than did either young or old adults (appendix C, table C.3). Annual variation in birthrate for young adults was four times that for intermediate-age females. Mean annual number of births recorded for the ninety-five adult females (including relocated animals) was 16.3 ± SE 1 calf per year, which equals an annual birthrate of 7.6 ± 0.8%.

MORTALITY. In a review of causes of population decline in bear species, Garshelis (in press) concluded that reducing human-induced adult mortality is the key to conservation and recovery. This statement applies to virtually all large mammals. Because they are long lived and relatively slow breeders, many large mammals will suffer severe declines in their populations if excessive hunting, poaching, or other sources of adult mortality are

*SD = standard deviation; SE = standard error.

not brought under control. I was keen to investigate what effect human-induced adult mortality had on the trajectory of the Chitwan population and of the populations in the reserves where rhinos had been translocated from Chitwan.

Natural mortality is, of course, a rare event for large, long-lived, and endangered herbivores. Researchers seldom learn of corpses quickly enough to determine cause of death. Luckily, Nepal considers rhinoceros the property of the king. As it is also a sacred animal, each death triggers a formal inquiry; therefore, data on mortality are generally quite accurate. In addition, congregations of vultures tip off guards, elephant drivers, researchers, and nature guides to a death.

Twenty-eight animals died in the Sauraha subpopulation between 1984 and 1988 (table 5.3). As one would expect for a long-lived giant herbivore, mortality for all age classes was low. The annual mortality rate for calves was $2.8 \pm 0.9\%$; for subadults, $2.2 \pm 0.7\%$; and for adults, $2.9 \pm 0.5\%$. All calf mortality occurred during the first year of life when calves were prey for tigers. Similar data from Kaziranga National Park in Assam show that tigers may take as many as 15% of all young born each year. Beyond this age, rhinoceros in both reserves, like most megaherbivores, are largely unaffected by predation. The disproportionate number of mortalities among male adults (15 of 18) suggests that competition for mates may be the most important contributor to deaths of males.

Yonzon (1994) reported sixty-four deaths between 1988 and 1994, an average of eleven per year. On average, five subadults or adults died from natural causes each year (usually from wounds suffered during fights), two calves as a result of tiger predation, and four from poaching. The two calves per year killed by tigers are probably an underestimate because young calves are not often seen until they are several weeks old. Unless one is closely following radio-collared females that have just given birth, mortality data on very young calves are difficult to obtain. Regardless, the crude annual mortality rate of 2.4% between 1988 and 1994, which Yonzon (1984) calculated for the Chitwan subpopulation, is very close to Laurie's (1978) estimates and mine.

Mortality from disease seems to be rare. Outbreaks of septicemia are known for Kaziranga, but epidemics have not been an issue in Chitwan or elsewhere to date. The potential for epidemics to devastate existing populations is quite a serious threat that has led to the design of translocation programs in Nepal and India. Reseeding parks that historically harbored rhi-

TABLE 5.3. NUMBERS AND CAUSES OF MORTALITIES FOR GREATER ONE-HORNED RHINOCEROS BY SEX AND AGE CLASS IN THE SAURAHA SUBPOPULATION OF ROYAL CHITWAN NATIONAL PARK, 1984–1988

	Adults						Subadults			Calves		
	6–<12 yr		12–<20 yr		>20 yr		4–<6 yr					
Cause of death	M	F	M	F	M	F	M	F	Sex unknown	0–<1yr	1–<4yr	Total
Poaching	—	—	2	—	—	—	—	1	1	—	—	4
Tiger predation	—	—	—	—	—	—	—	—	—	4	—	4
Intraspecific fighting	—	—	—	—	5	1	1	—	—	1	—	8
Separation from mother	—	—	—	—	—	—	—	—	—	1	—	1
Flood or quicksand	—	1	—	1	—	—	—	—	—	1	—	3
Cause undetermined	1	—	4	—	3	—	—	—	—	—	—	8
Total	1	1	6	1	8	1	1	1	1	7	0	28

noceros, and that are now well protected from poachers and contain suitable habitat (Mishra and Dinerstein 1987), will remove the extinction threat from a severe localized outbreak of disease.

Restoration of Rhinoceros in Their Former Range: Founder Populations in Bardia and Dudhwa National Parks and Sukla Phanta Wildlife Reserve

Chitwan served as the donor population for three separate translocation efforts geared toward reestablishing greater one-horned rhinoceros across their range. Four adult females were sent to Dudhwa National Park in Uttar Pradesh, India, near the border with Nepal, in 1986. Translocations to Bardia National Park occurred in 1986, 1990, 1999, 2000, 2001, and 2002 and involved a total of seventy animals. The sex and age composition of the founder population in Bardia was biased toward adult females. The first group of thirteen was established in the western part of Bardia near the Karnali River. The animals translocated in 1990 were released in the Babai Valley in the eastern part of the park, but individuals from these two translocations were able to mix. The Babai Valley has been the release site for all subsequent translocations (table 5.4).

By 1998, we could begin to assess the effectiveness of this translocation effort. Twenty-seven calves were born in Bardia since the translocations.

TABLE 5.4.   POPULATION SUMMARY FOR THE FOUNDER POPULATION IN ROYAL BARDIA NATIONAL PARK, DECEMBER 1986–NOVEMBER 2000

	Number	Remarks
Translocated from Chitwan (1986)	13	Released in Karnali floodplain
Translocated from Chitwan (1990)	25	Released in Babai Valley
Translocated from Chitwan (1999)	4	Released in Babai Valley
Translocated from Chitwan (2000)	16	Released in Babai Valley
Total translocated	58	
Calves born in Bardia	27	
Natural mortality	−11	
Poaching mortality	−7	
Estimated current population (2001)	65	

Eighteen deaths across all age categories have occurred. The annual loss from natural mortality is almost identical to that of the original Chitwan population. However, nine animals, all adults, died at the hands of poachers. Today the recorded net gain is only seven animals from the original fifty-eight translocated from 1986 to 2000. This fact underscores Garshelis's (in press) point about the dramatic effect that human-induced adult mortality, here attributable to poaching, can have on small populations. If rhinoceros are well protected, the population should begin to increase dramatically; the founder population has adjusted well in Bardia, and enough excess carrying capacity remains for many more rhinoceros to be added.

Rhinoceros were common in Dudhwa in the early nineteenth century but decreased steadily in response to overhunting. The last animal in the vicinity was shot in 1878 (Sinha and Sawarkar 1993). More than 100 years later, in 1984, two adult males and three adult females were translocated from Kaziranga to Dudhwa. Two females died soon thereafter, so an additional four females were translocated from Chitwan to Dudhwa in 1986. There has been some gain in the population, with numbers increasing from eleven in 1993 to thirteen in 1995 (Foose and van Strien 1997). As in Bardia, however, poaching has prevented the population from increasing as rapidly as expected. Perhaps additional translocations from Chitwan may bolster the size of the Dudhwa population.

Apparently, one adult male from Dudhwa escaped several years ago and eventually took up residence across the border in the Royal Sukla Phanta Wildlife Reserve in Nepal. In November 2000, Nepalese wildlife staff translocated three adult females and one adult male to Sukla Phanta. The presence of extensive lush riverine grasslands suggests that Sukla Phanta has great potential as a rhinoceros reserve if only poaching can be effectively controlled.

Genetic Threats of Extinction

The genetic consequences of low population numbers and brushes with extinction are a source of great concern in conservation (Allendorf 1986; Allendorf and Leary 1986; Frankel and Soulé 1981; Franklin 1980; Lande and Barrowclough 1987; Schonewald-Cox et al. 1983). Severe reductions in population size, known as bottlenecks, tend to diminish genetic diversity (typically expressed in terms of heterozygosity, allelic diversity, and polymorphism). A prolonged population bottleneck may lower the fitness of individuals and jeopardize the long-term survival and evolutionary potential of their populations (Allendorf 1986; Franklin 1980). Researchers have attributed the lack of genetic diversity in populations of various rare and endangered species to bottlenecks (Bonnell and Selander 1974; O'Brien and Evermann 1988; O'Brien et al. 1985, 1987; Pemberton and Smith 1985). Although the consequences of losing genetic diversity are serious for normally variable populations, scientists now question whether they have overemphasized the genetic effects of bottlenecks. For one thing, bottlenecks must be quite small and repeated or sustained for several generations for major erosion of heterozygosity (Allendorf 1986; Chakraborty, Fuerst, and Nei 1980; Lande and Barrowclough 1987; Nei, Maruyama, and Chakraborty 1975). Moreover, the probability of sudden extinction is high when population size is small (Goodman 1987; Lande 1988; Pimm et al. 1989). Thus the loss of genetic diversity because of bottlenecks may be less of a problem than current literature suggests because most small populations probably will become extinct before losing a substantial portion of their existing variability (Lande 1988; Pimm et al. 1989).

Because of the attention given to bottleneck effects, designers of recovery plans for wild species have focused on the preservation of genetic diversity (Lande 1988). Unfortunately, much of the data supporting the genetic consequences of small isolated populations comes from work on captive animals in zoos. Researchers are not likely to conduct pre- and postbottleneck genetic surveys for many wild populations, and the relationship between current genetic diversity and past demographic events will remain inferential in most cases. However, we can evaluate the strength of this inference by considering demographic features of populations—mating systems, dispersion, and dispersal patterns—that affect levels of variation and by obtaining the best data available on historic population sizes. For example, if we can estimate current and past effective population sizes (N_e)— roughly, the fraction of the population that actually breeds—we can calculate expected erosion rates of genetic diversity before the bottleneck (Lande and Barrowclough 1987). This then allows us to assess the plausibility of the bottleneck scenario for explaining the levels of diversity observed in current populations. Even approximate estimates of effective population size will allow us to make calculations that could be important to the inferential arguments on which this controversy centers.

High Heterozygosity in the Chitwan Population: The Effects of Historical Demography, Dispersal Capabilities, and Generation Time

The low numbers of rhinos in Dudhwa, Sukla Phanta, and Bardia and their relative isolation from Chitwan raise a couple of questions: How different is the management of these reserves from zoos, where biologists must guard against the erosion of genetic diversity? What happens if rhino numbers fail to increase quickly and the population endures a prolonged bottleneck?

Gary McCracken collaborated with me to evaluate data on the genetic diversity in the Chitwan population (Dinerstein and McCracken 1990). The results were startling: nine of the twenty-nine presumptive loci that we examined were polymorphic (appendix C, table C.4). The overall heterozygosity (H_o) measured from this suite of loci (Hedrick et al. 1986) was 9.9 ± 4.5%, among the highest observed in free-ranging mammals.

The heterozygosity that we documented in the Chitwan population is in stark contrast to the much lower levels that others have reported for populations of other species that have experienced near extinction (Bonnell and Selander 1974; O'Brien and Evermann 1988; O'Brien et al. 1985, 1987; Pemberton and Smith 1985). This observed heterozygosity is also at the extreme high end of values observed in the more than 140 other mammal species that have been examined using similar techniques (Wooten and Smith 1985). Two intriguing questions arise: How did such high levels of genetic variability accumulate in these large mammals before their reduction in numbers? And how has this variability persisted through the bottleneck that the Chitwan population experienced in the late 1950s and 1960s?

In the fifteenth century, before extensive human settlement within its range, rhinoceros inhabited floodplains, oxbows, and feeder streams of major rivers from northwestern Burma across the Indo-Gangetic Plain to the Indus River valley in northern Pakistan (Laurie 1978) (see figure 1.9). Left undisturbed, rhinoceros are, and probably were, abundant in their primary habitat—the tall grasslands found along major rivers. Average densities in Kaziranga National Park currently approach 4 individuals per km^2 (Foose and van Strien 1997). In Chitwan, as I mentioned earlier, peak seasonal densities reach 13.3 individuals per km^2. The area of historic prime habitat was approximately a 4-km-wide band along major rivers, resulting in 35,800 km^2, a conservative estimate of the area. This area, multiplied by current Chitwan densities in prime habitat (and excluding marginal habitat), gives us a minimal total population estimate of 476,140 individuals.

The ability of individuals to disperse, particularly over hostile terrain (e.g., human-dominated landscapes), heavily influences the chances for an endangered species to recover its more natural population level. Rhinoceros are highly mobile; even within their restricted current range, individuals move linear distances of more than 60 km within a year (Gagan Singh, personal communication, 1988). This tendency to disperse easily along the floodplains at the base of the Himalayas suggests that rhinos historically could move freely; they could circumvent human settlements by moving across and along floodplains or even across fields at night. This behavior is also observed in Asian elephants (Anil Marander, personal communication, 1998). Tigers, on the contrary, rarely cross agricultural lands if the distance between blocks of natural habitat exceeds 5 km (Dinerstein et al. 1997).

Thus the mobility of rhinoceros, coupled with their wide range across the Indian subcontinent and probable high density, all suggest that, around A.D. 1400, rhinoceros could easily have had effective population sizes of tens of thousands. High levels of genetic diversity can accumulate in populations of this magnitude, provided that they sustain their large effective size for many generations (Nei 1987; Soulé 1976). Fossils of greater one-horned rhinoceros date from the Middle Pleistocene, and fossils also demonstrate a broader prehistoric distribution for this species than is estimated for around 1400 (Laurie et al. 1983). Therefore, rhinoceros probably persisted in very large numbers for at least 100,000 rhinoceros generations. It also seems probable that much of the accumulated genetic diversity in this species was widely distributed throughout its range, with little or no population structuring among regions. The precise calculation of effective population size during this period or any other period is problematic because both the necessary life table statistics and full information on the variance in individual reproductive success are lacking (Lande and Barrowclough 1987).

Still, McCracken and I have sufficient information from field studies to approximate it. Of 251 Chitwan rhinoceros in 1984 to 1988, 87 were breeding females and 51 were breeding-age males (derived from appendix C, table C.2). Mature females produce one calf approximately every four years, and variance in female reproductive success appears to be low. Throughout the study period, only 28 of the 51 adult males showed evidence of breeding activity (chapter 7). The remaining 23 adult males never attained dominance, were not allowed to approach estrous females, and showed none of the behavioral and morphological characteristics obvious in breeding males. Assuming discrete generations and that the variance in progeny number equals the mean number produced per individual (excluding nonreproductives), we calculate that $N_e = 85$, or about 35% of the total population (Lande and Barrowclough 1987). Greater one-horned rhinoceros clearly violate both assumptions, but it is not yet possible to evaluate the net effects of these violations on effective population size. Therefore, we use $0.35 \times N$ (N = total population size) as the best estimate of N_e. These conclusions will not be qualitatively affected even if this estimate is incorrect by a factor of two (more on this shortly).

The effect of long generation time is to slow the erosion of genetic variability. The rate of decay of heterozygosity resulting from small population size is known to be approximate $1/(2\ N_e)$ per generation (Allendorf 1986; Lande and Barrowclough 1987). Our estimate of average generation time in free-ranging rhinoceros is about twelve years, which is lower than that calculated for other rhinocerotids (Thomas Foose, personal communication, 1988). Before 1950 Chitwan maintained a population of at least 1,000 individuals ($N_e < 350$), so the population should have lost no more than 6.4% of its original heterozygosity through the forty-six rhinoceros generations since a.d. 1400. After 1950 the rate of loss of heterozygosity would have accelerated. However, because only about three generations have elapsed since the population's precipitous decline, and because recovery has been rapid (i.e., in 1962, $N_e = 21-28$; in 1975, $N_e = 95$; in 1988, $N_e = 133$), we calculate that further erosion of heterozygosity probably did not exceed an additional 3%. Therefore, the current population in Chitwan should retain approximately 90% of the heterozygosity present when the species was still widespread and common. If effective population sizes were actually half our estimates, approximately 82% of the original heterozygosity should as yet be preserved, whereas if effective population sizes were twice our estimates, more than 95% should still be present. These estimates of the heterozygosity preserved are probably conservative because the Chitwan population undoubtedly exceeded 1,000 individuals between 1400 and recent times, and our estimate of length of generation time is probably too low. Although Chitwan rhinoceros retain high heterozygosity, we observed relatively low allelic diversity, with three alleles at two loci and two alleles at all other polymorphic loci. High allelic diversity is an expected result of sustained large effective population sizes (Chakraborty, Fuerst, and Nei 1980; Nei, Maruyama, and Chakraborty 1975), and loci with multiple alleles are common in studies of organisms with apparently large effective population sizes (e.g., Ayala et al. 1972; McCracken 1984). Many of these alleles are at low frequencies and contribute little to overall heterozygosity (Allendorf 1986; Chakraborty, Fuerst, and Nei 1980; Fuerst and Maruyama 1986; Lande and Barrowclough 1987; Nei, Maruyama, and Chakraborty 1975). Rare alleles are lost quickly during bottlenecks (Allendorf 1986; Fuerst and Maruyama 1986), and this could explain the relatively low allelic diversity observed. However, with a sample of twenty-three individuals, we expect to see only about 20% of all alleles at frequencies of 0.001–0.01 and about 70% of those at frequencies of 0.01–0.5. Therefore, small sample size may preclude our detecting any loss of allelic diversity resulting from reduced population size.

Finally, our results are in contrast to the only other published electrophoretic study of greater one-horned rhinoceros. Merenlender and colleagues (1989) report no observed variation among three individuals derived from the Kaziranga population. As the authors suggest, more individuals from that population must be examined to determine whether the Chitwan and Kaziranga populations vary in the amount of genetic variation present. New analyses, using polymerase chain reaction analysis, are under way using tissue samples from Chitwan animals.

We conclude that high heterozygosity persists in this population because the population size remained large before 1950, the genetic bottleneck occurred recently, and average generation time is long. The observation of high genetic variability in the Chitwan population was surprising initially and is in contrast to the results of several other studies of genetic variability in rare and endangered species. However, our results can be explained by considering the distributional history, demography, and life history parameters of these animals. This study illustrates the need for considering these parameters when appraising genetic effects of population bottlenecks in other species. The conclusion that a population carries low diversity because it has experienced one or more bottlenecks is inferential. A cause-and-effect relationship between low diversity and small population size has not been demonstrated; in the studies cited as evidence of this phenomenon, researchers conducted genetic analysis of all species only after their populations had been reduced. We can estimate present levels of variability, but low diversity at present does not indicate how much diversity a population had in the past.

Chitwan rhinoceros provide an example of a population that almost went extinct while still carrying high genetic diversity. Given the accelerating rate of extinction, other threatened species that were until recently common and widespread may yet retain a substantial proportion of their original heterozygosity. Another lesson for conservation biologists is that at the current rate of habitat loss, poaching, and other human influences, most small isolated populations of endangered large mammals will probably become extinct long before deleterious genetic effects could become manifest in the populations.

Genetic Considerations in Recovery of Endangered Large Mammals

Conservation biologists and managers may find themselves in the difficult position of being forced to cut corners in the restoration of endangered species. Restoration may require the mixing of populations that might be considered subspecies in order to rapidly build up numbers in founder populations. For example, Dudhwa National Park now has individuals from both the Assam region and Chitwan, two populations thought by some to show some degree of morphological separation (Groves 1993). My view is that conserving the ecological role of greater one-horned rhinoceros in natural systems (chapter 8) is more important than running the risk of maintaining the "genetic purity" of source and founder populations while watching those populations disappear forever.

Environmental Risks of Extinction

Even if a rhinoceros population recovers rapidly from a population bottleneck and retains much of its original heterozygosity, environmental risks of extinction pose another set of threats to recovery.

Designing appropriate recovery strategies requires knowledge of habitat–density relationships and carrying capacity for determining whether rhinos require large areas to maintain viable populations and how pristine their habitat must be. With this in mind, the staff of the King Mahendra Trust for Nature Conservation (KMTNC) and I undertook the most extensive radiotelemetry studies to date; the studies involved twenty-one individual rhinoceros for four years (appendix A). Supported by the staff of the KMTNC, this study produced a data set supplemented by additional information collected on nine radio-collared individuals in the Bardia founder population by Shant Raj Jnawali, a scientist at the KMTNC.

The highest rhinoceros densities in the Sauraha subpopulation of Chitwan coincided with census blocks containing large tracts of *Saccharum spontaneum* grassland, the forage plant that fecal analysis shows rhinoceros consume most frequently (Jnawali 1995) (table 5.5). In 1986 density in the Icharni block exceeded 10.5 rhinos per km^2. Thus compared with other rhinos, greater one-horned rhinoceros are clearly able to reach and sustain extraordinary densities in native habitats (table 5.6). In Chitwan rhinos also reached high densities in blocks covered largely by degraded scrub in the buffer zone (e.g., see the entries under Darampur and Badreni-Kharsar in table 5.5), even though these blocks were also adjacent to agriculture. Such densities were seasonal and coincided with the ripening of the rice crop in October and November.

Blocks covered mainly by *Narenga porphyracoma* grassland, the most common grassland association in Chitwan (Lehmkuhl 1988), supported lower densities (table 5.5). Rhinoceros avoided dense stands of *Themeda arundinacea* along the edge of *sal* forest in the Jarneli, Simalchaur-Kachwani, and Bansbari blocks. Blocks with the lowest densities (Simalchaur-Kachwani) lay farthest from the Rapti River and included the most *sal* forest. In other blocks, rhinoceros were concentrated near the Rapti River. Densities did not correlate with the size of census blocks.

Contrary to conventional wisdom, the highest densities of rhinoceros were not always on the border with agriculture. Densities in Dumria, Ghatgain–Kanu Taal, and Patch 3, Baghuwaghera — blocks in which resident rhinoceros fed almost exclusively on natural forage (Jnawali 1986) — exceeded or equaled densities in census blocks bordered by croplands. Rhinoceros in the West subpopulation also reached locally high densities around the Tiger Tops Lodge, more than 3 km from the edge of cultivation. Densities in blocks bordering agricultural areas seem to fluctuate seasonally in association with the maturation of rice, corn, wheat, and lentils. During the rice harvest in October 1987, densities in the Badreni-Kharsar block were 8.3 per km^2 and declined to 3 per km^2 by February 1988 after grass fires began inside the park. The tallgrass species of Chitwan respond quickly after burning. New shoots became abundant within two weeks after the fires, and most animals vacated the Badreni-Kharsar block to feed on the flush of new growth in *S. spontaneum* grasslands inside the park.

Densities of greater one-horned rhinoceros correlate positively with the percentage of the block covered by *S. spontaneum* grassland. Along

TABLE 5-5. GREATER ONE-HORNED RHINOCEROS DENSITIES AND SOME HABITAT FEATURES OF CENSUS BLOCKS IN ROYAL CHITWAN NATIONAL PARK, APRIL 1988

Area and block	Size (km^2)	Rhinoceros density (N/km^2)	Percentage of block covered by S. spontaneum grassland	Distance from croplands to center of highest density (km)	Dominant plant associations in the block
Sauraha					
Icharni	4.14	9.4	52.6	0.5	S. spontaneum, riverine forest
Darampur	4.16	2.2	0.0	0.5	Scrub
Dumria	3.62	9.4	29.6	5.0	Narenga grassland, S. spontaneum
Badreni-Kharsar	1.05	3.0	0.0	0.5	Riverine forest, scrub
Patch 1 and 2	2.41	13.3	43.6	1.0	Riverine forest, S. spontaneum
Patch 3, Baghuwaghera	3.15	12.4	29.2	3.0	Riverine forest, S. spontaneum
Simalchaur-Kachwani	4.76	1.7	0.0	1.0	Narenga grassland, sal forest
Bansbari	1.80	1.1	0.0	3.0	Narenga grassland, sal forest
Jarneli	1.96	3.6	0.0	3.0	Narenga grassland, sal forest
Ghatgain	3.49	8.3	5.6	3.0	Riverine forest, Narenga grassland
Subtotal	30.54	6.4	—	—	
Kagendramali	6.84	4.7	—	1.0	
West	60.18	1.2	—	3.0	
Bandarjhola and Narayani	13.33	3.0	—	3.0	

NOTE: Estimates do not include relocated animals and are period of fallow crops. For the locations of census blocks, see figure 5.2.

TABLE 5.6.   COMPARISON OF POPULATION DENSITIES OF GREATER
ONE-HORNED RHINOCEROS AND OTHER SPECIES IN SELECTED AREAS

Species	Location	Area (km²)	Year	Density (N/km²)
Greater one-horned rhinoceros	Chitwan, Sauraha population	30.5	1988	6.40
	Royal Chitwan National Park			
	Excluding buffer zone	932	1994	0.48
	Bardia, West population	70	1998	0.21
	Kaziranga	430	1995	2.79
	Dudhwa	490	1995	0.03
	Katarniaghat	20	1995	0.20
	Orang	76	1995	1.18
	Pabitora	18	1995	3.22
	Jaldhapara	216	1995	0.16
	Gorumara	79	1995	0.16
Sumatran rhinoceros	Gunung Leuser, Sumatra	1,400	1995	0.01
	Taman Negara, Peninsular Malaysia	4,400	1995	0.01
	Tabin, Sabah	1,200	1995	0.01
White rhinoceros	Greater Umfolozi complex, South Africa	940	1970	2.10
	Umfolozi Game Reserve, South Africa	480	1970	3.20
	Tsavo East, Kenya	544	—	1.60
Black rhinoceros	Hluhluwe, South Africa	215	1961	1.40
	Serengeti, Tanzania	—	—	0.04
	Mara Reserve, Kenya	—	—	0.14
	Etosha, Namibia	—	—	0.01

NOTE: The comparison uses the findings of Owen-Smith (1988) and Foose and van Strien (1997). Densities are low for some greater one-horned rhinoceros populations because of heavy poaching.

stream banks, *S. spontaneum* can account for more than 90% of aboveground biomass. Not surprisingly, *Saccharum* is a staple in rhinoceros diets; it normally exceeds 50% of the diet each month (Jnawali 1986). *Sacchrum spontaneum* is exceptional among the common tall perennial grasses of Chitwan in that new shoots sprout soon after cutting, grazing, burning, or inundation by floods. Most other species sprout only once per growing season, regardless of these disturbances. New shoots of *S. spontaneum* also provide nutritious forage. Nitrogen content in new growth is twice that of mature leaves and ten times that of stems (chapter 6).

Laurie (1978) claims that rhinoceros in Chitwan reach their highest densities in areas with the greatest vegetation diversity. He assessed vegetation diversity qualitatively by counting the number of different vegetation types in a given area. However, I present counterevidence that the highest densities do not relate to degree of vegetation diversity but to the abundance of *S. spontaneum*. This species forms nearly monospecific associations along river terraces within each block. The other common association in the high-density blocks is patches of riverine forest. This forest association also exhibits low within-habitat diversity where two species, *Trewia nudiflora* and *Ehretia elliptica,* dominate 77% of the canopy. However, densities of rhinoceros were always higher in blocks where *S. spontaneum* grasslands bordered riverbanks. The distribution of habitats used by rhinoceros is different in Bardia than in Chitwan. Nevertheless, Jnawali's (1995) study of nine radio-collared, translocated animals shows that, as in Chitwan, rhinoceros occur in highest densities in tall grassland during the hot season and monsoon season.

Monsoon floods are the most influential component of Chitwan's disturbance regime. During the flood of the Brahmaputra River valley in 1988, about 70% of Kaziranga National Park remained under 2.7 to 4 m of water for several weeks. Thirty-nine rhinoceros drowned in the rising waters ("Kaziranga Under Water" 1988). The monsoon season of 1998 led to another major flood in Kaziranga. The rhinoceros population faced high mortality because recent land clearing limited their access to upland forested areas above the floodplain of the Kaziranga sanctuary. When rhinoceros and other ungulates become isolated from these upland forests, floods can become a major cause of calf mortality. Thus upland forests in Chitwan, Bardia, and Kaziranga, which large herbivores mostly avoid for much of the year, serve as important refuge during high-water periods. Inclusion of relatively unproductive upland *sal* forests in Terai wildlife reserves is critical for the long-term survival of the rhinoceros populations in the area.

Answering Questions Related to Management of an Endangered Species

Four important observations made during the fieldwork have direct relevance to rhinoceros conservation and address the questions that I posed at the beginning of part II.

1. The data presented here show that greater one-horned rhinoceros reach high densities in the earliest successional habitats, *Saccharum spontaneum* grasslands. These habitats are maintained by a large-scale natural disturbance event—the annual monsoon floods. The high productivity of these habitats means that a large number of breeding individuals can occupy a small area, minimizing the spatial requirements for viable populations of this large mammal species.

2. Recovery of greater one-horned rhinoceros habitat may take only one year after a major disturbance event such as a flood or heavy overgrazing by domestic stock. Rhinoceros do well in degraded scrub forests as long as *S. spontaneum* grasslands are available close by. A problem for rhinoceros is that the areas they prefer are distributed as a narrow strip along river systems rather than deep in the interior of reserves. These areas are more likely to fall along the border of parks because many reserves use stream banks as boundary demarcations. Furthermore, they lie close to human settlements.

3. Not all grasslands are of equal value for rhinoceros conservation and management; rhinoceros essentially shun the tallest grasslands dominated by *Narenga porphyracoma,* which account for most of the grasslands in the park. Maximizing the area covered by *S. spontaneum,* in either the park or the buffer zone, is the best method for accelerating the recovery of rhinoceros populations.

4. High densities of rhinoceros are not dependent on proximity to croplands. However, because of foraging preference and thermoregulatory strategies (addressed in chapter 6), this species is restricted to living close to rivers.

Do Greater One-Horned Rhinoceros Qualify as an Extinction-Prone Species?

Terborgh and Winter (1980) claim that the features that tend to foster extinction are limited geographic range (rarity), large size, and habitat specialization. Greater one-horned rhinos are both large and habitat specialists, preferring early successional conditions (the first plants that colonize the floodplain). Are these the real reasons for considering rhinos to be extinction-prone species? In chapter 2, I explained how poaching for rhino horn overrides the natural ability of rhinos to recover quickly. Thus it is impossible to ignore the issue of poaching when determining whether the rhinoceros is prone to extinction.

But if modern rhinoceros lacked horns and Asian and African elephants lacked tusks — thereby removing the incentive for most of the intensive poaching pressure — would the ecological correlates of very large body size still place them in jeopardy of extinction? I believe that the answer is no, based on the rapidity with which these populations recover where they are well protected from human disturbance in productive habitats. Another near megavertebrate — the wild Asiatic water buffalo (*Bubalis bubalis*) — helps shed light on this question. The horns of wild water buffalo, although quite impressive, are of no commercial value. Yet, at least in Nepal, this species is even more endangered than greater one-horned rhinoceros. The reason for its endangerment is that, like the rhinoceros, its preferred habitat is the subtropical and tropical floodplain of Asia. Unfortunately, both the rhino and the water buffalo are in competition with the subcontinent's subsistence rice farmers, who also consider the floodplain to be prime habitat.

Thus I propose the addition of a new criterion to Terborgh and Winter's (1980) list of extinction-prone features. If a species prefers highly productive habitats suitable for intensive agriculture, such as most subtropical or tropical floodplains, it should be considered highly extinction prone. This preference for a particular habitat also in demand by humans far outweighs the effect of any demographic, genetic, or behavioral correlate of extinction risk. In addition, having body parts to which humans attribute high economic value is enough to bring an animal to the brink of extinction or beyond.

Statistical Notes

Sex and Age Structure and Breeding Status

1. Intermediate-age adult females (12–20 years) were significantly more numerous than males ($X^2 = 4.3, p < 0.05$).
2. These differences in age structure for the populations in 1975 and 1988 largely result from Laurie's (1978) designation of subadults as

3 to 9 years old, whereas we placed them between 4 and 6 years. If we reclassify the 1988 data into the same categories that Laurie used, the age structure is the same in 1975 and 1988.

Pattern of Calf Births

1. We could detect no significant difference in the distribution of births during the calendar year ($X^2 = 2.3$, d.f. $= 11$, $p < 0.99$).
2. We arranged our calf birth data in the same bimonthly format that Laurie (1978) used for his study in 1972 to 1975; the combined data set ($N = 113$) revealed no seasonality to parturition, either ($X^2 = 8$, d.f. $= 5$, $p < 0.10$).

Genetic Analysis

1. Genotypic frequencies at each polymorphic locus conformed to Hardy-Weinberg expectations (Levene 1949).

Habitat–Density Relationships

1. The highest rhinoceros densities in the Sauraha area of Chitwan coincided with blocks that include large tracts of *Saccharum spontaneum* grassland ($r = 0.7750$, $N = 9$, $p < 0.05$).
2. Densities did not correlate with the size of blocks ($r = 0.3333$, $N = 9$, $p < 0.20$).
3. High densities are not related to proximity to agriculture ($r = 0.0297$, $N = 9$, $p < 0.50$).

CHAPTER 6

LIFE ON THE FLOODPLAIN:
SPACING AND RANGING BEHAVIOR,
FEEDING ECOLOGY, AND ACTIVITY PATTERNS

In Asia, special attention must be given to those species of ungulates which appear to exploit early to mid-succession vegetation stages. These species are especially vulnerable to decline in small parks and reserves where vegetation succession can move rapidly to a climax condition.
—John F. Eisenberg and John Seidensticker, "Ungulates in Southern Asia"

I SAW MY FIRST LIVING greater one-horned rhinoceros in the company of three other Peace Corps volunteers on a sweltering afternoon in May 1975. We had been sent to Chitwan as part of our training program but with scant introduction to the wildlife of the park or warnings about the dangers of walking in the jungle. Naively, we tried to get a close-up photograph of a male rhino immersed in a wallow. The male charged without warning, chasing us into the nearest climbable tree. For the next few minutes, while the male snorted and circled menacingly below us, I had the chance to observe him from our unusual vantage point. Although I was fresh out of college, then and there I planned future studies, the results of which I detail in this chapter.

In Chitwan, as in any natural habitat, the daily routine of a wild ungulate consists of feeding, resting, thermoregulating, and avoiding predators. Is the rhino's lifestyle fundamentally different from that of other floodplain grazers? Does its size have ecological consequences? The effects of body size among vertebrates have been of major interest to ecologists, physiologists, and, recently, conservation biologists. Reviews by Calder (1984) and Schmidt-Nielsen (1984) focused largely on the physiological aspects of large animal ecology. Other recent studies have examined the ecological correlates of body size (Peters 1983), particularly for megaherbivores (Owen-Smith 1988).

"What relevance do these species' characteristics have for the wildlife manager?" wondered an East African game warden who had come to Nepal in 1975 to help establish Chitwan and Bardia National Parks. "If you are trying to save large mammals," he bluntly asked, "why not simply set aside the biggest area you can, walk the perimeter, gazette [establish] the boundaries, and form a dedicated guard patrol to keep the poachers out? Who needs to know how long a rhinoceros spends feeding or wallowing or where it goes during the monsoon?"

As it turns out, such information is essential for managing endangered species. Body size influences the size of the home range, seasonal and daily movements, diet selection, and activity patterns, and a conservation strategy for rhinoceros requires consideration of these features. We collected the data on rhinoceros in their prime habitat. Would the results be similar in an area that contains less prime habitat? What happens when rhinos are translocated to less optimal sites, or at the edge of their range, where populations are likely to be under greater stress than in the core of their historical distribution? We needed to hypothesize how translocated populations would behave, move about, and feed relative to the Chitwan source population. To answer these and other questions, I compared space use and feeding ecology between the source population of rhinoceros in Chitwan and the translocated population in Bardia. This study relied on twenty-four-hour watches of a substantial number of habituated or nearly habituated radio-collared rhinoceros. I used the data from these individuals to determine the time they spent feeding and to detect differences in activity patterns related to season and sex. I also focused on wallowing behavior — an activity that varies seasonally but is crucial to the survival of rhinoceros on a hot and humid floodplain.

Spacing Behavior, Home Range, and Daily Movements

The small size of Asian reserves presents problems for area-limited species. The problem is acute for highly territorial species that occur at low densities, such as tigers. For example, maintaining a breeding population of 500 tigers would require a space that exceeds the size of all three of Nepal's major tiger sanctuaries combined. Megaherbivores that undergo extensive seasonal migrations between habitats, such as wild elephant and Indian bison, also require sufficiently large areas set aside to maintain viable populations for the long term.

Greater one-horned rhinoceros differ markedly in spacing behavior when compared with these other large mammals that share their habitats. In Chitwan, and likely in Kaziranga, rhinos occupy what are probably the smallest annual ranges, seasonal home ranges, and core use areas recorded among megaherbivores. Annual mean home-range size among dominant adult breeding males was 4.3 km^2 and among adult breeding females was 3.5 km^2 (table 6.1). Three females had young calves during most of the monitoring period; the size of their annual home range was less than half that of the other two females, which were unaccompanied by calves. The two males that were dominant in the highest-density female area occupied mean annual home ranges and core areas of the same size as the ranges of breeding females (table 6.1). These two males occupied home ranges that did not overlap, but the range of one (M009) was twice as large as the range of the other (M008) (figure 6.1).

The size of home range varied seasonally for females and among individual females within the same season (appendix D, table D.1). Home-range size reached a maximum in the hot season (4.6 km^2) and contracted to 2.3 km^2 during the monsoon season when forage was abundant. Two females occupied core use areas (i.e., 50% of all locations) of 22 and 28 ha, respectively, during the monsoon season. These were remarkably small for such a large grazer — and ten times smaller than that previously reported for this species (Owen-Smith 1988). In contrast, home-range size among dominant males expanded slightly during the monsoon season to reach the seasonal maximum of 4.1 km^2.

Individual adult females overlapped considerably in seasonal and annual home ranges and therefore are not illustrated. High overlap and constant home-range size attest to a high level of home-range fidelity. In contrast, dominant males occupied temporally or spatially distinct home ranges. Two other males overlapped in area but maintained dominance at different times. When one replaced the other as the dominant male on Icharni Island, the home range of the nondominant male constricted to the edge of the area used by the dominant male (M038).

Adult females rarely traveled more than 5 km during a twenty-four-

TABLE 6.1. ANNUAL HOME-RANGE SIZES OF GREATER ONE-HORNED RHINOCEROS IN ROYAL CHITWAN NATIONAL PARK

| Sex | Number of locations | Area (km²) | | | Percentage of locations in *Sa. sp.* | Percentage of locations in core area home range in *Sa. sp.* | Percentage of locations in home range in RF |
		Core area 50% of locations	Annual home range 95% of locations	Minimum convex hull			
Adult females							
Gajuri F010	147	0.58	2.06	2.62	47.0	53.7	29.3
Khagchiruwa F002	104	0.47	2.59	2.72	73.4	66.3	32.7
Tindharke F004	91	0.71	5.63	7.12	70.0	65.9	23.1
Laxmi F014	68	0.85	5.21	5.58	35.0	51.5	42.6
Abire Pothi F003	101	0.33	1.84	2.98	76.7	60.4	38.6
Mean ± SD		0.6 ± 0.2	3.5 ± 1.8	4.2 ± 2.0	60.4 ± 18.4	59.6 ± 6.8	33.3 ± 7.7
Breeding adult males							
Conan M038	86	0.56	3.64	4.74	25.0	50.0	43.0
Karne M008	101	0.59	3.22	4.21	57.7	48.5	46.5
Yadav M009	80	1.82	6.10	7.64	13.9	22.5	15.0
Mean ± SD		1.0 ± 0.7	4.3 ± 1.6	5.5 ± 1.8	32.2 ± 22.8	40.3 ± 15.5	34.9 ± 17.3

NOTE: Core areas represent 50% of all locations; home range is 95% of all locations. Both calculations were derived using harmonic surface isopleth method (Ecological Consulting, Inc. 1993); convex hull method is included for comparison. Minimum convex hull is a standard technique for estimating home range in wildlife field studies. It involves determining the smallest area that would contain a given number of observations as determined by the researcher. *Sa. sp*, *Saccharum spontaneum* grasslands; RF, riverine forest.

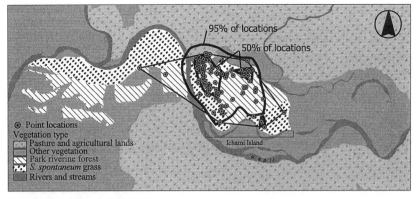

Home range of Karne (M008)

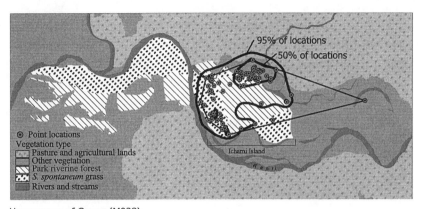

Home range of Conan (M038)

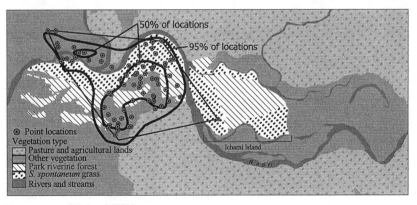

Home range of Yadav (M009)

FIGURE 6.1 Home ranges of three dominant male rhinoceros in Royal Chitwan National Park. Each figure illustrates core areas (50% of all locations) and home range (95% of locations). M008 and M038 were dominant on Icharni Island, but during different time periods. The Rapti River separated the home range of M009 from Icharni Island.

hour period in any season (appendix D, table D.2). Adult breeding males traveled slightly longer distances during the twenty-four-hour cycle. These movements were sufficiently large to allow animals to cover most of the core area in a few days. Rarely during any of the observation periods did individuals venture farther than 2 km from the river's edge.

Five years after translocation from Chitwan, the annual home range of radio-collared individual animals in Bardia averaged 25.1 km^2 for adult females and 41.8 km^2 for adult males (Jnawali 1995) (table 6.2). Jnawali (1995) used a minimum convex polygon method to estimate home ranges, so I recalculated the home-range data in my study by the same method to enable comparisons. Bardia rhinoceros used annual home ranges more than eight times larger than those of radio-collared individuals in Chitwan. Seasonal home ranges among females in Bardia were three to seven times larger than for females in Chitwan. The difference among males was even more dramatic: Bardia males occupied seasonal home ranges seven to ten times larger than those of breeding males in Chitwan.

Why are the home-range sizes of Chitwan rhinoceros so small? The simplest explanation is that rhinos in Chitwan occupy prime habitat — the lush, dense floodplain dominated by *Saccharum spontaneum* grasslands. This habitat is far less extensive in Bardia (Dinerstein 1979; Jnawali 1995). About 60% of core areas and annual home-range polygons occupied by adult females in Chitwan consisted of *S. spontaneum* grasslands (table 6.1). The water table in the Chitwan floodplain is also higher; the grasslands of Chitwan maintain a soil moisture content of 20 to 30% throughout the year (Lehmkuhl 1988). This condition favors the maintenance of *S. spontaneum* grasslands. In contrast, the substrate of the Geruwa River floodplain in Bardia consists of a layer of young alluvial sandy soils over a thick layer of

TABLE 6.2.   COMPARISON OF ANNUAL AND SEASONAL HOME RANGES (KM²) FOR RADIO-COLLARED GREATER ONE-HORNED RHINOCEROS IN ROYAL BARDIA AND ROYAL CHITWAN NATIONAL PARKS

	Females		Males	
	Bardia	Chitwan	Bardia	Chitwan
Cool	15.6 ± 4.5	2.4 ± 0.7	20.7 ± 10.3	2.9 ± 1.8
Hot	13.3 ± 4.9	2.7 ± 1.7	21.2 ± 11.1	2.7 ± 0.8
Monsoon	14.4 ± 7.1	2.4 ± 0.9	14.4 ± 7.1	3.7 ± 2.0
All year	25.1 ± 9.3	2.9 ± 0.9	41.8 ± 4.4	3.3

boulders. In Bardia's hot season, soil moisture content drops below 5% (Lehmkuhl 1989). Lack of soil moisture probably prevents resprouting of *S. spontaneum* in Bardia until very late in the hot season, whereas in Chitwan this species sprouts all year long.

Another explanation is that translocated animals tend to wander rather widely. Subadults often wander farther than adults, and this age group constituted the bulk of animals translocated to Bardia. However, Jnawali did not start his telemetry study in Bardia until 1991, five years after the translocation. Presumably, these animals would have settled into smaller home ranges if the distribution of *S. spontaneum* grasslands was more expansive. Perhaps the highly skewed sex ratio in Jnawali's population (1 breeding male per 8 females) and much lower densities (0.3 animal per km^2 in Bardia versus 8–10 animals per km^2 in Chitwan) contributed to increased movement among Bardia rhinoceros and explain the larger seasonal and annual home ranges.

Comparisons with Other Herbivores

The home-range sizes of the two other species of Asian rhinoceros are not readily comparable because telemetry studies are lacking. Borner (1979) estimated the home-range size for Sumatran rhinoceros to be as large as 50 km^2, based on recognizable footprints of distinct individuals. Habitat type also influences the annual home-range size of black rhinoceros, a browser like Sumatran and Javan rhinoceros. Female black rhinoceros in the Lerai Forest of the Ngorongoro Crater in Tanzania use home ranges of less than 2.6 km^2 but expand their range to 99 km^2 in the Serengeti grasslands where browse is presumably less available year round (Frame 1980; Goddard 1967). More typical home-range sizes across southern and eastern African habitats are 7 to 35 km^2. The largest home ranges recorded for all rhinoceros species are those of the desert-dwelling black rhinoceros in Namibia, whose range is larger than 500 km^2 (Owen-Smith 1988). In the best-studied population of white rhinoceros in the Umfolozi Reserve, Owen-Smith (1988) estimated annual home range for females at 9 to 15 km^2. In contrast, Owen-Smith estimated the home ranges of adult males, which are territorial, to be 0.7 to 2.6 km^2. The largest Asian megaherbivores, Asian elephants, occupy home ranges of about 85 km^2 in southern Sri Lanka (Global Environment Facility 1999). In the Nilgiri Hills of southern India, two family herds used home ranges of about 105 km^2 and 115 km^2 each. Home ranges of three identified solitary adult bulls were 170 km^2, 320 km^2, and 215 km^2 (Sukumar 1999).

How does the relationship between home-range size and movements of greater one-horned rhinoceros compare with that for smaller ungulates on the floodplain? For example, are home ranges of 1,800-kg rhinoceros fifty times larger than those of the 35-kg hog deer that share their habitat? The answer is no. Home-range size for male hog deer and adult female hog deer averaged 0.8 km² and 0.6 km², respectively (Dhungel and O'Gara 1991), only about one-half to one-third the size of the smallest home ranges of female rhinos in Chitwan.

In sum, the home-range analyses revealed that greater one-horned rhinoceros use remarkably small annual and seasonal home ranges in prime habitat and that their ranges are not scaled to body size. This finding, from an ecological perspective, suggests that, at least in the spatial context, rhinoceros behave more like a small grazing ungulate than a megaherbivore. Given the high range overlap among females, and small home-range size, I have concluded that rhinoceros sanctuaries that contain at least 400 km² of prime habitat are likely to support a minimum viable population of 1,000 rhinoceros over the long term. However, if translocation programs for rhinos consist of sites with lower levels of prime habitat (*S. spontaneum* floodplain), as in Bardia, we can expect that home ranges will expand considerably and that rhinos will need much larger areas to maintain a viable population.

Feeding Ecology

Both astute field biologists and successful poachers have detailed knowledge of what, where, and when their target eats. Most ungulates devote a considerable part of the day to finding, ingesting, and — in the case of true ruminants with a four-chambered stomach — reingesting plant material. Understanding the feeding ecology of a particular species is fundamental to formulating a conservation strategy for it. Is the species a diet generalist or specialist? Does it have to wander far to find adequate forage? Metabolic requirements and type and physiology of the digestive tract strongly influence food choice among herbivores.

Larger animals generally eat more food per day than do smaller ones. However, megaherbivores require fewer nutrients per unit ingested than do smaller ungulates that have higher metabolic rates. We should expect Chitwan's megaherbivores — rhinoceros, Indian bison, and elephants — to eat more food and spend more time foraging than hog deer or hispid hares, although the latter two species should eat more nutritious food. Hog deer differ markedly from rhinoceros not only in size but also in gut morphology. Hog deer, like axis deer, sambar deer, swamp deer, and bison, are true ruminants. Their four-chambered stomachs facilitate digestion of plant parts high in cellulose because the animal can remasticate the ingesta. The major advantage of the ruminant digestive system is that remastication increases the surface area of ingested forage for bacterial action.

Hispid hare, horses, and rhinoceros are cecal digesters. To compensate for their less efficient digestive tracts, nonruminant cecal digesters have grown bigger. Large body size and a high ratio of gut size to body size automatically confer an advantage in digesting structural cellulose through bacterial fermentation because the food remains in the gut for so long (Owen-Smith 1988). This is an alternative way of enhancing cell wall digestion in the absence of a four-chambered stomach. Still, the digestive efficiency achieved through large size cannot compare with that attained by certain grazing ruminants. Grazing rhinoceros increased their digestive efficiency by developing a huge digestive sac, the cecum, which enables them to break down plant tissues as hind-gut fermenters.

Chitwan's Most Important Grassland as Viewed by a Large Grazer

The enormous quantity of green-plant matter that covers Chitwan's grasslands suggests a paradise for grazing ungulates. Indeed, the grass there is more plentiful than almost anywhere else on Earth. Ironically, most plant communities in Chitwan are composed of vegetation high in indigestible material or slowly digestible plant parts that are relatively low in nutritional content. For example, consider forage availability and quality of Chitwan's *S. spontaneum* grassland, rhinos' favorite forage (figure 6.2). New growth, the shoots most nutritious for ungulates, accounts for only 0.3% of the total available plant material at the beginning of the hot season. The least nutritious components, green stems and dry leaves and stems, contribute 96% of available plant tissues (table 6.3). Thus lush, easily digestible forage is scarce. Megaherbivores in Chitwan that rely on highly digestible forage face a serious challenge to ingest enough to meet their metabolic needs. One solution is to evolve food-gathering structures to compensate for the limitations imposed by the high fiber–low nitrogen content found in Chitwan grasslands. Greater one-horned rhinos have evolved a prehensile upper lip. Expansion of lips is a phenomenon seen in other grazing megaherbivores, such as

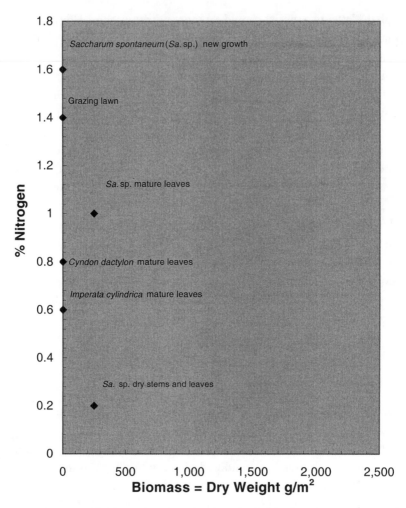

FIGURE 6.2 Availability and nutritional quality of typical *Saccharum spontaneum* grassland, Royal Chitwan National Park.

white rhinoceros and hippos, and browsers such as black and Javan rhinos. This adaptation allows greater efficiency in plucking nutritious forage.

Another factor that influences feeding length is the distribution of the most heavily used forage. If an individual is able to establish a home range in prime *S. spontaneum* grassland, not only is foraging area reduced but probably also time spent foraging. Another adaptation for a large-bodied herbivore is to store fat, particularly when the quality and abundance of forage fluctuate widely. White rhinoceros build up reserves of subcutaneous fat

TABLE 6.3.   BIOMASS OF VARIOUS PARTS OF *SACCHARUM SPONTANEUM* AND
OTHER SPECIES IN CONTROL *S. SPONTANEUM* GRASSLANDS SAMPLED IN ROYAL
CHITWAN NATIONAL PARK, JANUARY 1987

Plant species and part (N = 8 plots)	$\overline{X}$ (grams dry weight/m²)	Standard deviation (SD)	Percentage of total
S. spontaneum			
New growth	5.90	4.46	0.003
Last 25 cm of mature leaves	15.79	3.87	0.008
Other green leaves	43.40	17.08	0.02
Green stems	274.79	108.70	0.13
Dry leaves and stems	1,716.80	582.12	0.83
Subtotal *S. spontaneum*	2,077.24	690.01	0.99
Other species	0.40	0.40	0.0002
Total biomass	2,078.55	690.50	

to enhance survival through the dry season (Owen-Smith 1988). Although no one has noted indications of fat reserves in carcasses of greater one-horned rhinoceros, it remains an area for future research.

Researchers have hypothesized that in places like Chitwan, nonruminants should outperform ruminants on forages of high fiber content because indigestible material does not hamper their food passage rate (Bell 1971; Foose 1982; Janis 1976). However, for greater one-horned rhinoceros to survive, they need to obtain sufficient quantities of digestible energy despite the high passage rate of plant material (known as the throughput rate) through their digestive tract. A few questions arise: How different are the diets of rhinos from those of smaller ungulates? Among rhinoceros populations, are there significant differences in diet selection between populations in prime habitat and those that have less access to prime habitat? The discussion that follows draws heavily on two published works, Jnawali's (1995) comparison of food habits of rhinoceros in Chitwan and Bardia using fecal analysis of rhino dung containing plant tissues, and Dhungel and O'Gara's (1991) study of hog deer.

Diet Selection

For greater one-horned rhinoceros, grass species constituted about 86% of the diet during the hot season in Chitwan and roughly 92% of the total diet during the monsoon season in Bardia. In the cool season in both areas, consumption of grass species drops to a seasonal low at 57% in Chitwan and 42% in Bardia. In both areas, Jnawali (1995) found that rhinoceros consumed agricultural crops more commonly during the cool season (13%) and seldom during the monsoon season (5%). The proportion of browse in the diet was higher in Bardia than in Chitwan except during the monsoon season (figure 6.3), when rhinoceros ate large amounts of *Trewia* fruits, a species uncommon in Bardia. The highest ratios of browse to grass were recorded during the cool season in both areas (figure 6.4).

Jnawali (1995) recorded 283 plant species available to rhinos in Chitwan and 179 plants species in Bardia. The highest numbers were recorded in tall grasslands (131), followed by riverine forest (117). In Bardia the highest number of species was observed in riverine forest (93). Not surprisingly, the diet of rhinoceros was diverse, but fewer than 10 species contributed more than 75% of the total volume in the diet in both areas (table 6.4). In Chitwan four grasses (*Saccharum spontaneum, Saccharum benghalensis, Cynodon dactylon,* and *Narenga porphyracoma*) and three browse plants (*Coffea benghalensis, Murraya paniculata,* and *Litsea monopetala*) contributed more than 85% of the annual diet. Similarly, in Bardia five grasses (*Saccharum spontaneum, Arundo donax, Cynodon dactylon, Saccharum benghalensis,* and *Erianthus ravennae*) and four browse species (*Mallotus philippinensis, Dalbergia sissoo, Callicarpa macrophylla,* and *Calamus tenuis*) accounted for about 75% of the volume of the annual diet.

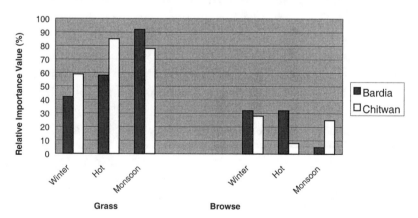

FIGURE 6.3 Diet of greater one-horned rhinoceros in Royal Chitwan and Royal Bardia National Parks. (Adapted from Jnawali 1995:68, fig. 4)

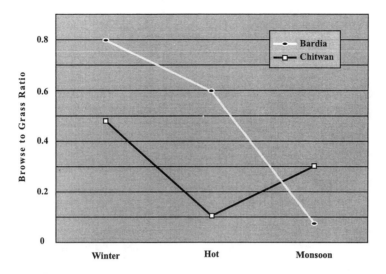

FIGURE 6.4 Seasonal variation in the ratio of browse to grass in the diet of rhinoceros. Symbols represent mean ± SD. (Adapted from Jnawali 1995:69, fig. 5)

The proportion of plant species eaten in each season varied slightly, but rhinoceros in Bardia consistently consumed a larger proportion of available plants than did the rhinos of Chitwan. Rhinoceros ate nearly 30% of the total number of different wild plant species available in Bardia and about 13% of those available in Chitwan (figure 6.5). In Bardia rhinos ate the highest proportion of available plants (24%) during the cool season, whereas in Chitwan they consumed the highest proportion (11%) during the monsoon season.

Saccharum spontaneum was by far the most important food plant during all seasons in both study areas (table 6.4). In Chitwan this species contributed most to the diet during the hot season, whereas in Bardia rhinos consumed it most often during the monsoon season. It was least important during the cool season in both areas. Among important browse species, rhinos in both study areas preferred *Mallotus philippinensis* during the cool season (figure 6.6). In Chitwan, Jnawali (1995) recorded the highest preference of browse species for *Dalbergia sissoo* and *Coffea benghalensis*.

In sum, rhinos selectively ate certain species in each season, and rhinos in both areas exploited food plants in proportion to their availability. Stud-

TABLE 6.4. RELATIVE IMPORTANCE VALUES OF MOST COMMON WILD FOOD PLANTS IN THE RHINOCEROS DIET, ROYAL BARDIA AND ROYAL CHITWAN NATIONAL PARKS

| Species | Relative importance value | | | | | | | |
| | Cool | | Hot | | Monsoon | | All year | |
	RB	RC	RB	RC	RB	RC	RB	RC
Grasses								
Saccharum spontaneum	18.9	25.7	21.2	43.1	45.4	41.9	28.5	36.9
Saccharum benghalensis	0.8	14.9	3.2	13.8	8.7	8.2	4.2	12.3
Narenga porphyrocoma	—	1.6	0.7	8.4	2.3	8.4	1.0	6.1
Erianthus ravennae	2.1	—	3.8	—	4.7	—	3.5	—
Cynodon dactylon	4.4	4.3	4.7	7.6	3.1	8.2	4.1	6.7
Imperata cylindrica	—	0.4	4.4	2.3	1.2	2.6	1.9	1.8
Themeda sp.	—	3.1	—	2.2	—	2.8	—	2.7
Cymbopogon sp.	0.7	2.3	2.0	3.2	3.8	0.5	2.2	2.0
Phragmites karka	1.9	0.7	1.5	1.2	2.2	0.8	1.9	0.9
Arundo donax	5.6	—	5.4	—	4.5	—	5.2	—
Browse								
Callicarpa macrophylla	3.9	3.7	4.5	1.0	3.2	2.0	3.9	2.2
Litsea monopetala	—	5.0	—	0.4	—	0.6	—	2.0
Coffea benghalensis	—	6.5	—	2.8	—	3.0	—	4.1
Murraya paniculata	—	5.8	—	2.1	—	4.0	—	3.9
Mallotus philippinensis	7.9	2.6	5.9	0.3	0.6	0.4	4.8	1.1
Dalbergia sissoo	—	—	7.9	—	0.7	—	2.9	—
Trewia nudiflora	—	0.2	—	—	0.1	11.2	0.03	3.8
Calamus tenuis	4.4	—	5.0	—	0.9	—	3.4	—
Bombax ceiba	1.2	0.2	0.6	0.1	—	—	0.6	0.1
Colebrookia oppositifolia	1.6	0.1	0.8	—	0.1	0.2	0.8	0.1
Ehretia elliptica	1.1	—	0.3	—	0.1	0.2	0.5	0.1
Ficus racemosa	1.7	—	0.1	—	0.3	—	0.7	—
Ziziphus mauritiana	1.0	—	0.1	—	—	—	0.4	—
Acacia concinna	1.3	0.1	0.4	—	0.2	—	0.6	0.03
Other								
Triumfetta sp.	0.4	0.4	0.1	—	0.1	0.2	0.2	0.2
Urena lobata	0.9	0.1	1.8	0.1	0.6	0.1	1.1	0.1
Circium wallichii	4.2	0.1	3.1	1.5	1.3	—	2.9	0.5

NOTE: Relative importance value is calculated as the mean percent of species X in fecal sample multiplied by the square root of the frequency of species X in the fecal sample. RB, Royal Bardia; RC, Royal Chitwan.

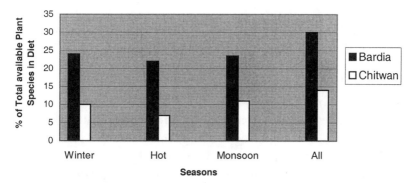

FIGURE 6.5 Proportion of total available plant species in the diet of rhinoceros in two study areas. (Adapted from Jnawali 1995):70, fig. 6)

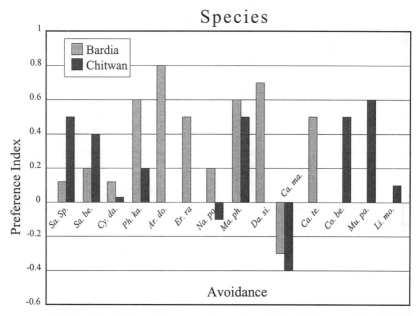

FIGURE 6.6 Electivity indexes of important forage plants in Royal Chitwan and Royal Bardia National Parks; *Sa. sp., Saccharum spontaneum; Sa. be., Saccharum bengalensis; Cy. da., Cynodon dactylon; Ph. ka., Phragmites karka; Ar. do., Arundo donax; Er. ra., Erianthus ravennae; Na. po., Narenga porphyracoma; Ma. ph., Mallotus philippinensis; Da. si., Dalbergia sissoo; Ca. ma., Callicarpa macrophylla; Ca. te., Calamus tenuis; Co. be., Coffea bengalensis; Mu. pa., Murraya paniculata; Li. mo., Litsea monopetala.* (Adapted from Jnawali 1995:72, fig. 8)

ies on axis deer (Mishra 1982) and on hog deer (Dhungel and O'Gara 1991), combined with my own observations of these species and of sambar, show that rhinos eat many of the same species but in much greater quantities, particularly in *S. spontaneum* grasslands. The dominance of *S. spontaneum* in this vegetation type ensures that any ungulate that lives in it must use this species for forage. Hog deer and sambar also feed on many of the same seasonally abundant fruits and certain fleshy flowers (*Bombax ceiba*) that rhinos eat. Observations by Dhungel and O'Gara (1991) and those from this study suggest that hog deer seek out younger shoots of grasses more consistently than do rhinoceros.

From a conservation perspective, we now know that, in translocated populations of rhinoceros where prime habitat is less extensive, the diet tends to be more diverse but that rhinos still concentrate on a few grass and browse species. Therefore, it follows that sufficient alluvial grasslands containing large areas of *S. spontaneum* or other *Saccharum* species must be available for the population to meet its nutritional requirements. It is necessary to survey the extent and abundance of such grasslands as well as determine the phenology of *Saccharum spontaneum* before initiating translocation programs. When this species becomes dry during the cool season, and presumably less palatable, adjacent riverine forests containing key browse plants must be available or rhinoceros will seek out agricultural crops. Overlap in species of plants selected by large and small herbivores on Chitwan's floodplain suggests that management strategies to benefit large ungulates benefit smaller endangered species — hog deer, swamp deer, and hispid hare — as well.

Activity Patterns

Comparing data for the activities of megaherbivores and wild ungulates puts the behavior of greater one-horned rhinoceros into context and sheds light on three issues. To what extent are rhinos diurnal or nocturnal feeders? How do seasonal changes in climate influence feeding and other activities, particularly wallowing? To what extent do adult males differ from adult females in activity patterns?

Where poaching is a dominant threat, wildlife managers need to know when animals are active, specifically moving and feeding, and therefore more easily exploited by poachers. A particular problem for megaherbivores is that their high overall metabolic requirements force them to eat both night and day. When big grazers feed out in the open in the late afternoon, they become easy targets for poachers. Clever poachers sometimes wait for rhinos to head to or depart from wallows. The design of poaching patrols would benefit from knowing where and when a species is active or seasonally nocturnal. Differences in male and female activity may lead to a differential probability of poaching mortality if one sex is more active or more visible at a certain time of day than the other. Finally, it is important to understand whether a particular activity, such as wallowing, is conducted only at certain sites. This knowledge can contribute to the design of effective translocation programs by ensuring that wildlife managers take certain actions, such as protecting wallow sites and oxbows and providing salt licks.

Foraging Activity

Owen-Smith (1988) distinguishes two types of feeding activity. Feeding is the act of browsing a piece of leaf or grazing on a grass blade; foraging includes the act of grazing or browsing plus movements of animals while seeking food.

During the cool season, rhinos exhibit about five small grazing peaks during the day, whereas browsing occurs in two pronounced peaks (figure 6.7). During the hot-dry season, grazing activity is distributed bimodally, and time spent browsing is minimal during the twenty-four-hour cycle. Grazing activity in the monsoon season exhibits the sharpest bimodal peaks. Intensive foraging and other activities require that animals rest periodically. Resting shows a daily peak early in the cool season, later in the hot season, and no clear pattern in the monsoon season.

When the feeding times of megaherbivores are different for each sex, females generally feed for longer periods (Owen-Smith 1988). In greater one-horned rhinoceros, this is clearly the case, as females graze for longer periods in each of the three seasons (table 6.5). Females still forage longer than males except during the hot season, when males forage slightly longer than females.

Our twenty-four-hour year-round observations of twenty-eight females revealed that females spend about 36% of their time feeding (figure 6.8), a figure significantly lower than that reported for female white rhinoceros. Male greater one-horned rhinoceros spend on average 28% of their time feeding. In the cool season, both sexes spend the highest percentage of

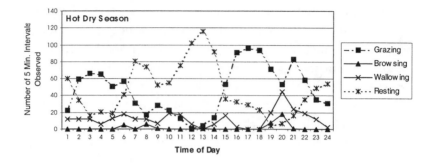

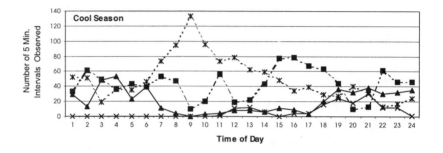

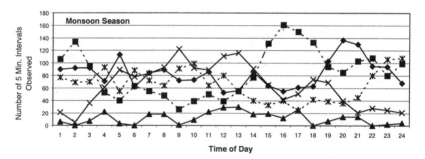

FIGURE 6.7 Daily activity patterns of greater one-horned rhinoceros at different seasons, based on twenty-four-hour activity watches of radio-collared, habituated individuals.

their day browsing, perhaps because they require additional time to move among patches where browsing occurs and must go farther to harvest browse plants than grasses.

Why do adult females spend more time foraging than do adult males? The simplest explanation of the Chitwan data is that most adult females under observation were pregnant, were lactating, or had just completed lactating. We would expect that the higher nutritional demands placed on fe-

TABLE 6.5.   FORAGING TIME BY SEASON AND SEX FOR GREATER
ONE-HORNED RHINOCEROS

| | Percentage of day | | |
	Foraging	Browsing	Grazing
Female			
Cool	52.7	14.5	27.3
Hot	52.6	1.1	34.3
Monsoon	50.6	4.4	32.3
All year	52.0	6.7	31.3
Male			
Cool	42.3	6.9	22.4
Hot	55.1	1.2	21.9
Monsoon	44.7	3.6	25.1
All year	47.4	3.9	23.1
Both sexes			
Cool	47.5	10.7	24.8
Hot	53.9	1.2	28.1
Monsoon	47.7	4.0	28.7
All year	49.7	5.3	27.2

NOTE: Foraging includes walking while browsing or grazing, whereas the columns under browsing and grazing represent the actual feeding time.

males by pregnancy or lactation would result in more time spent foraging. The males observed were typically dominant. Combat among males for control of feeding areas used by breeding females is so intense that males were probably spending their energy keeping track of females and warding off other males, leaving less time for foraging.

Thermoregulation

Ungulates native to tropical and subtropical environments must avoid overheating. The simplest solution is to forage at night and rest during the day. Unfortunately, high overall metabolic requirements, food-handling time, and digestion rates force megaherbivores to forage both day and night. This creates a major physical challenge — how to keep cool while feeding in the hot sun. Both rhinoceros and elephants show physiological adaptations to promote heat dissipation, such as large sweat glands in white rhinoceros (Cave and Allbrook 1958) and heat-radiating ear pinnae in African elephants (Buss and Estes 1971). Hiley (1977) and Langman (1985) suggest

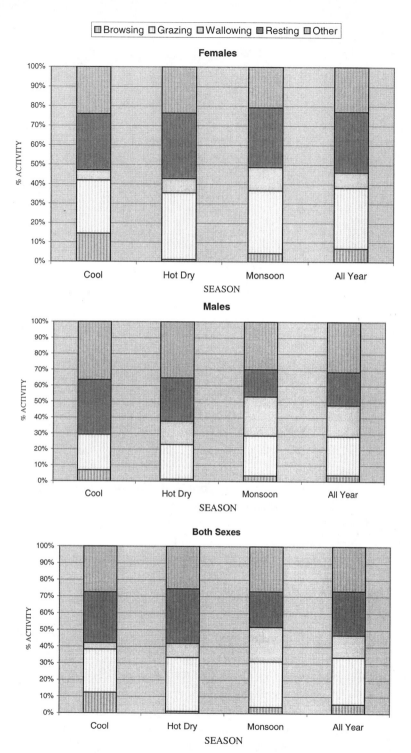

FIGURE 6.8 Activity patterns of adult female, adult male, and adult female and male rhinoceros, based five-minute scan samples from radio-collared, habituated individuals.

that the two African rhino species are able to store heat in their body tissues. Normal rectal temperatures for captive greater one-horned rhinoceros in Nepal are between 36 and 39°C throughout the year (Dinerstein, Shrestha, and Mishra 1990), so this may not be a strategy for this species.

Wallowing—immersing the body in water, often combined with rolling in mud—is widespread among large mammalian herbivores (Owen-Smith 1988). Wallowing's primary function is assumed to be heat regulation. Escape from biting insects as well as social communication by scent-marking may also encourage this behavior. For giant (1,000 kg) tropical herbivores living in open habitats, wallowing may be part of a suite of behavioral adjustments to heat stress that also includes nocturnal foraging and selection of dense shade during the daytime (Owen-Smith 1988).

Greater one-horned rhinoceros are known to use wallows (Laurie 1978) (figure 6.9), but detailed studies of heat regulation in this species do not exist. I observed that rhinoceros, during the mid- to late afternoon in the hottest but driest month of the year (April), grazed extensively and used wallows infrequently. During hot humid periods, conversely, rhinoceros spend much of the day in wallows, presumably to avoid heat stress. Several factors probably contribute to increased heat stress during the monsoon

FIGURE 6.9 Greater one-horned rhinoceros share a grassland wallow. (Photo by Eric Dinerstein)

season. Days are longer in the monsoon period, and thus total solar radiation is greater. Wind speed is also reduced in comparison with the other months of the year, so heat loss through evaporative cooling is less. Most important, the high concentration of water vapor reduces the ability of large-bodied herbivores to use evaporative cooling for heat stress.

I hypothesize that wallowing behavior is correlated with changes in vapor pressure density, a measure of the ability of air to hold water vapor at different temperatures. Wallowing behavior is related to other environmental variables and changes seasonally by habitat and by time of day. Heat stress, for example, restricts rhinoceros to riverine habitats during the monsoon season. This is, of course, an important finding for management of this endangered species and for consideration in future translocation efforts.

One major river, the Rapti, and several smaller streams flow through the 10.5-km^2 area where I studied wallowing behavior. This area contains more than forty major wallows (including old oxbows of the Rapti) scattered among tall grassland and riverine forest associations. More than sixty individuals frequented the wallows within the study area, and rhinos reached their highest densities in grasslands within 1 km of surface water (Dinerstein and Price 1991).

Some basic climatic data are helpful in interpreting patterns in wallowing behavior (figure 6.10). Mean maximum ambient temperature shows slight variation between March and October (i.e., hot season through the monsoon). The mean percentage of humidity remains relatively low from late February through early May, rises during the monsoon season, and remains high through the cool season. In contrast, vapor density pressure is high from June to October (monsoon) and drops substantially during other months. Most of the 2,000 mm of precipitation falls during the monsoon season (mid-May–mid-October).

Wallowing time was longest during the very humid monsoon period when it occurs around the clock. Rhinoceros wallow infrequently or not at all between November and February. Wallowing time, as expressed by the mean wallowing hours per day, correlated most strongly with vapor pressure density, precipitation, and mean maximum monthly temperature (see the statistical notes at the end of this chapter). Rhinoceros frequently use riverine forest wallows during the monsoon season and streams or rivers in the hot season. Riverine forest wallows provide more shade than do wallows in tall (3–7 m) grassland. The use of rivers and streams for wallowing from February through April is likely related to the drying up of favored riverine

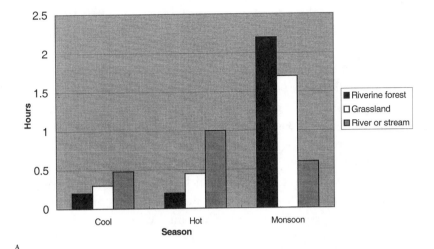

A

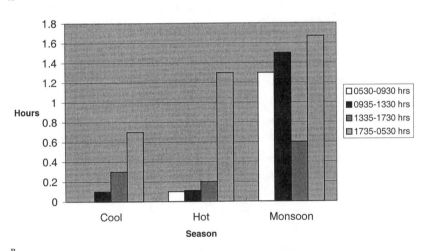

B

FIGURE 6.10 Wallowing habits of greater one-horned rhinoceros in Royal Chitwan National Park. (A) Mean wallowing hours per day by habitat type and season. (B) Mean wallowing hours per day by time period and season. The data are adjusted to account for unequal lengths of diurnal and nocturnal time intervals.

forest and grassland wallows during the hot season. May is a transition month when premonsoon rains often replenish grassland wallows. The low use of rivers and streams during the monsoon season is likely related to the force of the current, which makes wallowing less safe for females accompanied by small calves. Rhinos wallow most frequently at night during the hot-dry season and the cool season.

Wallowing might serve several functions with respect to thermoregulation: (1) to transfer heat directly to the cool mud or standing water, both of which are cooled through evaporation; (2) to cake mud on the body surface to reflect solar radiation; and (3) to evaporate wallow water directly from the skin. We did not measure these variables, but the data from Chitwan suggest that the combined effects of high temperature and high humidity, as measured by vapor density, largely drive seasonal changes in wallowing behavior. The relationship between water vapor concentration and ambient temperature is not linear; thus the hotter the temperature, the greater the amount of moisture the air can hold, which explains why vapor density is a better predictor of wallowing behavior than either mean maximum temperature or mean percentage of relative humidity. This phenomenon constitutes a heavy burden on a giant herbivore living in a humid, subtropical, or tropical environment. Conversely, wallowing is infrequent in the cool season despite the high relative humidity because temperatures remain low. High humidity in the cool season is attributable to a drenching mist that prevails each morning until about 10:00 A.M. The mist is probably sufficient to cool animals after extensive nocturnal feeding.

Wallowing is relatively infrequent during the hot-dry season and particularly during the day, because grazing rhinoceros reduce heat by sweating during the hot part of the day. Greater one-horned rhinoceros, like white rhinoceros, are capable of sweating profusely, as I saw when we were trying to catch them for translocation. The extent to which greater one-horned rhinoceros can store heat in body tissues has not yet been determined. During the monsoon season, the large mass and high volume-to-surface ratio of greater one-horned rhinoceros inhibits rapid heat loss through evaporative cooling, a mechanism observed in smaller ungulates (Taylor 1970). Long periods of immersion in wallows, often as much as eight hours per day, undoubtedly help reduce solar load and increase the amount of heat shed by conduction.

Biting insects and social communication are also cited as reasons for wallowing (Laurie 1978). However, biting flies (Tabanus spp.) are common during March and April when animals use wallows infrequently. Regarding communication, animals frequently scent-mark by urinating in wallows but did not defecate in a wallow during any of the observation periods. Rhinoceros defecate one to three times a day, and most defecations and frequent urination occur at latrines (chapter 8). It seems, then, that rhinos do not use wallows primarily for marking. Rhinoceros breed throughout the year, so the use of wallows as scent-posts to communicate estrus may play a role. Wallows as sites of social gathering probably explain why dominant adult males wallow for longer periods than do adult females with calves during two seasons of the year. Alternatively, females with calves may be forced to wallow less than males because calves need to forage more frequently. However, the observation that females wallow during the cool season but males do not challenges this theory, given that males do breed females during this season.

Heat stress largely restricts rhinoceros to floodplain habitats during the monsoon season. The susceptibility of riverine forests and grasslands to flooding during the monsoon season favors formation of wallows in early successional habitats versus the upland *sal* forest that covers much of Chitwan and seldom is affected by flooding. Most of the preferred forage occurs in the same early successional habitats; thus nutritional constraints may limit the distribution of rhinoceros in winter, with thermoregulatory constraints limiting the distribution of rhinoceros during the remainder of the year.

The implications of the wallowing study are obvious. Plans to reestablish rhinoceros populations within their former historic range (Khan 1989) should include a sufficient number of wallows as an important criterion in site selection for translocation efforts. Bardia has a much lower density of wallows and oxbows than Chitwan, particularly in alluvial habitats. This may be an important cause of the difference in behavior between translocated rhinoceros in Bardia and the original Chitwan population. Park managers can create and maintain wallows relatively easily. In areas where wallows are less common, management interventions can improve this feature for the benefit of rhinoceros and other species.

Statistical Notes

STATISTICAL RELATIONSHIPS AMONG CLIMATIC VARIABLES AND
WALLOWING BEHAVIOR

Wallowing vs.	Spearman r	p
1. Vapor pressure density	0.811	<0.002
2. Monthly precipitation	0.796	<0.002
3. Monthly mean maximum temperature	0.614°C	<0.005
4. Monthly mean percentage of relative humidity	0.049	ns

CHAPTER 7

MALE DOMINANCE, REPRODUCTIVE SUCCESS, AND THE "INCISOR-SIZE HYPOTHESIS"

The lifetime breeding success of males is most variable in species where direct conflict over mating access is common.
—T. H. Clutton-Brock, ed., *Reproductive Success*

ON A STEAMY JULY EVENING during my first monsoon season in Chitwan, I had just sat down to dinner when I heard tremendous crashing sounds in the riverine forest next to my camp. Galloping across the lawn at full tilt were two male rhinos, one chasing the other. Ignoring us completely, the males thundered past the dining area and straight through a barbed-wire fence, snapping the strands like party ribbon. This unexpected commotion right in the middle of my camp lawn initiated my investigation into the link between male dominance and reproductive success. The relationship between these is fully relevant to conservation, especially when wildlife populations dwindle to low levels. We need to look no further than zoo populations to recognize that, in small isolated populations of polygynous mammals, one or a few males can monopolize access to reproductive females. My first thought was that among Chitwan rhinos, a single powerful male would probably maintain dominance during the entire length of my intensive field study in the areas where females concentrated.

My field notes and Laurie's (1978) earlier study confirmed that males engage in fierce, sometimes fatal, combat to establish dominance and gain access to estrous females. Later on, as we learned to recognize individual rhinoceros, my staff and I observed older, once-dominant males hanging close to our elephant stable, where dominant male rhinos rarely wandered. The males that clung close to our camp were often badly scarred or limping heavily. For the defeated males, life on Chitwan's floodplain is, to paraphrase Thomas Hobbes, often nasty and brutish, and sometimes short.

Both Laurie (1978) and I observed that large sharp lower incisors rather than the horn inflicted serious wounds (see figure B4.3). Rather than repeat his excellent ethological study of rhinoceros, I focused my attention on the role that the condition of the incisors might play in intraspecific competition and reproductive behavior (for a concise ethogram for greater one-horned rhinoceros, see appendix E). Permission by the Nepal Department of National Parks and Wildlife to capture, immobilize, and measure rhinoceros on a regular basis provided a new opportunity to extend Laurie's work and examine breeding behavior more closely. In the first month, I noticed breakage and wear of both lower incisors on an old breeding male that we had immobilized to treat wounds inflicted by a challenging male. During the course of the four-year field study, we immobilized a large number of males to assess dominance (chapter 4). The dentition of old males that had died after fights showed heavy wear or breakage of mandibular incisors. These incisors are rootless and grow back, although probably quite slowly. Damage to incisors probably occurs because these teeth are more brittle than bone, because they are under compression, and because rhino males clash with great force. Examination of these teeth led me to formulate the "incisor-size hypothesis": the size and condition of the tusks (mandibular incisors) in breeding-age rhinoceros help determine dominance, access to estrous females, and ultimately reproductive success.

Is the hypothesis valid? Could it really be that in a 2,000-kg megaherbivore, the difference of a few centimeters in incisor length determines who gets to breed? A direct test of the hypothesis requires long-term studies of reproductive success in dominant males and removal of incisors from experimental animals. Such an experiment is logistically and ethically not possible because this population is still highly endangered. Instead, I explored the validity of the incisor-size hypothesis through field observations of changes in incisor condition, male dominance, and access to females. I examined the extent to which incisor length and other characteristics vary in association with male distribution and access to reproductive females. Most of my findings come from observations in an intensive study area in the east-central part of Chitwan and of the group of rhinoceros that lived in or adjacent to Icharni Island, a 3.2-km^2 island in the floodplain of the Rapti River.

Social Organization, Dominance, and Spacing Among Males

Breeding-age males (figure 7.1) compete to maintain status on Icharni Island. They attack other males that intrude on female congregations at forest wallows or in grasslands. Male rhinoceros are among the most aggressive of ungulates (Dinerstein, Shrestha, and Mishra 1988). Male–male combat accounts for nearly half of all mortalities recorded. Breeding males are solitary, but both Laurie (1978) and I occasionally observed two or three subadult or young adult males foraging or resting together within the home range of a dominant male. Telemetry studies and twenty-four-hour activity watches conducted on Icharni indicate that during certain periods of the

PRADHAN — M002

FIGURE 7.1 Portraits of the dominant males of Royal Chitwan National Park: M002, Pradhan; M008, Karne; M038, Conan; M039, Bhudo Bhaale; M031, Jawaani Karne; M016, Kankatuwa; and M009, Yadav. (Photos by Eric Dinerstein)

KARNE — M008

CONAN — M038

FIGURE CONTINUED

BHUDO BHAALE — M039

JAWAANI KARNE — M031

KANKATUWA — M016

YADAV—M009

year dominant males rarely leave the Icharni area except to feed in nearby croplands for a few hours each night.

My initial hunch — that one male would dominate breeding activity during my intensive field study on Icharni — could not have been more wrong. Between 1983 and 1989, seven males achieved dominance (figure 7.2). Length of dominance on Icharni averaged 7.9 months (SD = 4.7) and ranged from 1 to 15 months. The adult male that remained dominant the longest — from February 1984 to February 1985 — was the only adult male on Icharni at the time. Typically, one dominant male and one or two subdominant males were present simultaneously on Icharni. In five of the six turnovers, fighting continued for an extended period after an initial confrontation until the change was finalized. The mean length of postdominant subordinate status was 9 months (SD = 5.0; range = 0–16 months). Several previously dominant males were shunted off to the island's periphery. Here they lived in forest patches close to extensive human activity where we seldom saw the dominant male. While one male was dominant from late 1986 until early 1987, two others lived in a forest tract east of the Dhungre River by our research camp, and a fourth lived in the Badreni-Kharsar forest. Meanwhile, a fifth formerly dominant male remained along the northern scrub forest bordering Icharni.

Young adult breeding-age males avoid Icharni; they appear only as transients, and dominant males promptly chase them away. During most of 1984, when a sole male was resident, the thirteen younger males living across the Rapti River never attempted to establish themselves on the island.

Social Organization, Distribution, and Reproduction in Females

Females are solitary and are accompanied by their calves until the calves reach age four. Females with calves graze near other females or share wallows but do not associate in any consistent pattern.

In all, about twenty-eight adult female rhinoceros were living on or at least visiting Icharni during the four-year period. Eight were resident on Icharni during the entire study, and four more were resident until 1986 when they were translocated to other reserves; eight other females were transients. In addition, four were present on Icharni during the calculated dates of conception, and four adult females were rarely sighted on Icharni and not present during conception (appendix F, table F.1). Thus densities of breeding-age females on Icharni varied annually. The highest density recorded during this study (4.2/km^2) occurred from 1985 until early 1986,

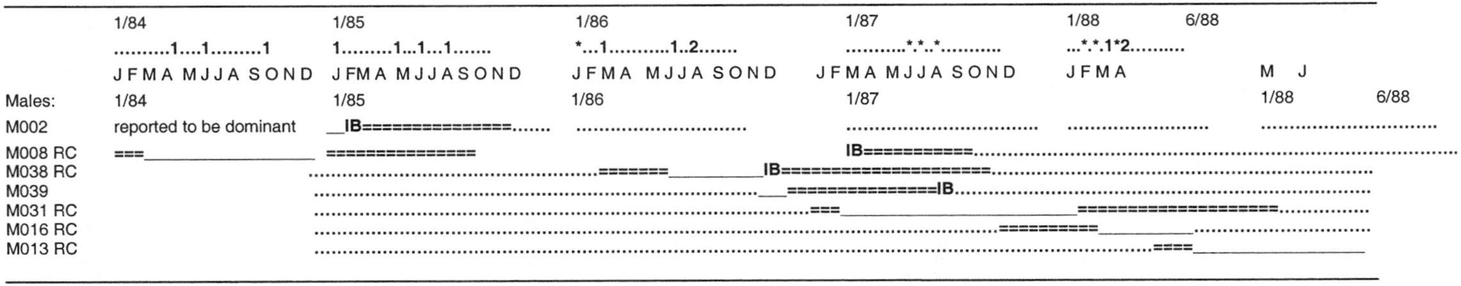

FIGURE 7.2 Periods of male dominance and conception dates for greater one-horned rhinoceros on Icharni Island, 1984–1988. The numbers above the abbreviations for months refer to the number of females that conceived during that month. Asterisks indicate that a female was mounted by a male, tended by the male, observed in estrus, or expected to be breeding, given the age of her last calf; RC, male radio-collared during all or part of his tenure on Icharni; IB, first discovery of broken lower incisor(s) in a male; ==, most dominant male; .., codominant; —, not present; —, most dominant male; ==, codominant.

just before the first translocation to Bardia. In contrast, densities of females in scrub and riverine forest or forest–grassland mosaics adjacent to Icharni contained far fewer females. The Badreni-Kharsar scrub contained 1.9 breeding-age females per km^2, the Darampur-Janakpur scrub contained 0.1 breeding-age females per km^2, and the Patch 1 area (riverine forest) held 0.8 breeding-age females per km^2.

Greater one-horned rhinoceros have one of the lowest reproductive rates among terrestrial mammals (chapter 5). A major reason is that free-ranging females are six years old before they start breeding. Why is breeding so delayed in females in the wild when females in zoos breed as young as four? One reason may be that females need to reach a certain body size to be able to withstand the rigors of courtship by the most aggressive dominant males. On two occasions males killed females they were chasing, and males sometimes wounded females during these chases. Long gestation periods and interbirth intervals also lower the reproductive rate: gestation lasts 15.7 months, and we calculated the mean interbirth interval as 45.6 months (SD = 6.4; range = 34–51) (chapter 5). Fewer opportunities for copulation and breeding probably heighten competition among males.

Reproductive histories of resident and transient female rhinoceros show that all but one conceived during the study (appendix F, table F.1). Six females were believed to have conceived twice, and three females may have conceived three times. The three females that conceived three times had shorter-than-normal interbirth intervals because their calves were removed for export to a captive-breeding program or were killed by a tiger.

Within an 11.5-km^2 study area, I recorded 21 known conceptions and estimated another 29 conceptions to have occurred during a 5.5-year interval. Based on data from known conceptions between January 1983 and January 1987, the mean number of conceptions for all females is 4.3 a year (SD = 0.4). Including data on estimated conceptions, the mean number of conceptions between January 1983 and January 1988 was 4.4 a year (SD = 1.1). Neither estimate accounts for prenatal or neonatal mortality.

Male Dominance and Estimate of Reproductive Success

The male that remained dominant longer than any other during the study (M008) did not experience an increase in breeding opportunities. Another (M031) was dominant for nearly half the period as M008, and more females came into estrus during this shorter interval (figure 7.1). A third male died

TABLE 7.1.   ESTIMATED NUMBER OF BREEDING OPPORTUNITIES FOR ADULT MALE GREATER ONE-HORNED RHINOCEROS ON ICHARNI ISLAND, ROYAL CHITWAN NATIONAL PARK, JANUARY 1, 1984–JUNE 6, 1988

	Individual male						
	M002	M008	M038	M039	M031	M016	M013
Estimated period as dominant male on Icharni (days)	30	450	150	30[b]	240	210	240[c]
Number of females estimated to have been bred while male was dominant	4[a]	4	2	0	3	3	6
Length of period of codominance on Icharni	270	510	360	180	360	90	60
Number of females estimated to have been receptive while male was codominant	3	4–5	4	4	4	0	2
Maximum estimate of total number of females bred (assuming all receptive females bred by this male while he was dominant and codominant on Icharni)	4	9–10	5–6	0	7	4	9

[a] Four females were receptive between January 1983 and January 1984 and probably were bred by M002.

[b] M039 died before he was able to impregnate any females on Icharni. The one copulation observed did not result in the pregnancy of F007.

[c] M013 was still resident and likely dominant on Icharni as of February 1990.

from internal injuries suffered in fights only thirty days after he assumed dominance. This male (M039) was observed copulating, but pregnancy did not result (table 7.1). He was attacked during this copulation by another challenger (M538), which, after driving off M039, mounted and copulated with the same female. Despite the high density of breeding-age females on Icharni, the long interbirth interval in rhinoceros yields few breeding opportunities (figure 7.3) during the tenure of any one male. The average minimum number of females assumed to be bred per male between January 1983 and June 1988 while each male was dominant was 3.3 (SD = 1.83).

Dominance and Physical Characteristics of Males

DOMINANCE AND BODY SIZE. In the field, older males are clearly distinguished from younger males by body size (height, girth, and length) (table

FIGURE 7.3   M031 and F010 copulate in riverine forest on Icharni Island. Note the radio collar on the female.

7.2). Breeding status among older males was again associated with body mass, extensive neck folds, elongated lower incisors, aggressive behavior, and squirt urination. Because data were lacking on body mass for each male, I used head and body length, neck circumference behind head, and neck circumference in front of shoulder as overall indicators of body size. Using these criteria, four of the seven breeding males on Icharni were among the largest males measured in the entire park. However, smaller males are still able to dominate larger males, as this study shows.

DOMINANCE AND HORN SIZE. Challenging males with shorter horns were able to evict four of the six dominant males on Icharni (table 7.2). Old males and females commonly have broken or eroded horns, and many females on Icharni had longer horns than did the males that bred them. Males could remain dominant even after their horns had broken. Within Chitwan as a whole, I observed six dominant males with broken horns.

DOMINANCE AND INCISOR SIZE. Incisors are razor sharp, and length is significantly dimorphic, reaching 9 cm in males. As teeth grow, incisors increase in width up to 4 cm at the base. Young adult males lack the dental weaponry found in older males.

TABLE 7.2. SOME MENSURAL CHARACTERISTICS OF GREATER ONE-HORNED RHINOCEROS, ROYAL CHITWAN NATIONAL PARK

| Character | Dominant males on Icharni, 1984–1988 | | | | | | | Adult males | | | | | | Adult females | | |
| | M002 | M008 | M038 | M039 | M016 | M031 | M013 | >12 yr | | | 6–12 yr | | | >12 yr | | |
								Mean	SD	N	Mean	SD	N	Mean	SD	N
Length of lower left incisor[a]	2.0IB	8.7	0.5 IB	3.0IB[b]	7.7	7.7	8.3	5.4	2.70	16	1.9	1.65	3	4.2	0.68	9
Length of lower right incisor	3.0IB	4.4	6.3	2.0IB[b]	8.3	6.1	8.0	5.6	2.44	14	1.9	1.65	3	4.4	0.77	9
Horn length	40	28	25	—	23	34	24	25.3	4.69	15	18	3.56	3	23.8	4.59	9
Horn circumference at base	—	—	63	—	48	53	54	59	11.95	11	46	1.63	3	46.2	3.05	9
Total body length	—	431	432	—	404	440	417	411.7	20.59	15	385	13.96	4	399.2	24.66	9
Tail length	—	70	63	—	70	70	66	65.7	5.13	15	63.8	2.28	4	63.9	9.99	9
Head and body length	—	361	369	—	334	370	351	346	18.39	15	321.3	12.32	4	335.3	25.71	9
Hind foot length to base of middle hoof	—	51	50	—	48	45	50	49.3	2.68	14	48.3	0.94	3	47.7	4.13	9
Ear length	—	25	23	—	24	25	25	24.7	0.85	15	25.5	0.50	4	25.9	1.66	9
Maximum skull circumference	—	183	173	—	164	163	180	170.7	10.51	10	148.5	12.13	4	157	8.22	9
Chest circumference	—	—	—	—	276	—	300	314.5	28.05	4	301	3	2	298.7	41.52	9
Neck circumference behind head	—	166	178	—	143	176	176	159.2	12.56	13	143	19.61	3	135.7	5.85	9
Neck circumference in front of shoulder	—	221	228	—	203	215	215	201.3	16.74	13	153.3	13.12	3	173	8.50	9
Shoulder width	—	60	65	—	59	61	58	67.2	11.55	11	47	6	2	59.7	3.95	9

NOTE: The tally includes breeding males on Icharni, adult males older than 12, adult males aged 6 to 12, and adult females older than 12.

[a] Length of incisors is from first capture date and is not necessarily indicative of incisor length immediately after assuming dominance.

[b] Measured at death. IB, incisor broken.

While dominant, males on Icharni had the largest incisors among all males measured in Chitwan. Loss in dominance in four males coincided with the breakage of one or both procumbent incisors during combat. In each known case, males with broken incisors remained on Icharni for various periods but eventually became reestablished elsewhere. Examination of incisor length and condition in all males upon assumption of dominance showed that both incisors were intact and equal to or longer than those of the defeated male.

In Patch 1, an area separated from Icharni Island by the Rapti River, males with the longest incisors also dominate other males (for the location of Patch 1, see figure 5.2). A male that retained dominance for more than three years (M009) had incisors measuring 8 and 7.2 cm in length (table

TABLE 7.3.   LENGTH OF PROCUMBANT LOWER INCISORS OF GREATER ONE-HORNED RHINOCEROS LIVING IN PATCH 1, ROYAL CHITWAN NATIONAL PARK

Number	Lower left (cm)	Lower right (cm)
M009	8.0	7.2[a]
M009	6.8	5.9[b]
M021	6.5	6.3[c]
Mean	7.1	28.06
SD	0.8	0.7
M114	0.7	0.7
M011	4.0	—
M012	4.5	4.5
M010	1.2	—
M001	0.9	1.0
M042	4.2	4.2
Mean	2.6	2.6
SD	1.7	1.8
Old adult males		
M024	0.5	0.5
M018	3.1	6.0
M007	5.1	6.0
M035	2.0	—
Mean	2.7	4.2
SD	1.7	2.6

[a] While dominant.

[b] No longer dominant.

[c] After M021 challenged M009.

7.3). This male dominated all others during this interval. All the other males had short or broken incisors. The home range of the dominant male extended west along the Rapti from Patch 1, probably because the part of Patch 1 across from Icharni contained few females. Younger males with smaller incisors lived close to the edge of Patch 1 along the riverbank. One adult male died from wounds inflicted by this dominant male. However, by mid-1987 the dominant male was challenged and eventually replaced (by M021). Recapture of the ousted male revealed a broken horn and damage to as well as wear of the incisors such that they equaled the length of the incisors of his successor.

Female Densities and Distribution of Males with Small Incisors

Within the study area, males possessing small incisors (i.e., 4 cm) were either young adults or previously dominant males with broken incisors. In general, males with small incisors live in forested areas containing few breeding-age females. For example, the Badreni-Kharsar scrub typically contains one older male with broken incisors and two or three younger transient males. The Darampur section of the study area also contains low densities of females and mostly younger males. Young adult males that use Patch 1 have peglike incisors, and older adults show attrition or breakage of incisor teeth. Throughout Chitwan, young adult and older males tend to congregate in outlying forested or scrub areas. In these areas, females are scarce in comparison with densities observed in the *Saccharum spontaneum* grasslands.

The movements of one male (M013) during the four years of study illustrate the changes in dominance, home range, and, ultimately, access to females in relation to development of incisor length. In 1984 this male lived in Patch 1 and was often sighted with a group of younger males. By early 1985, this male had moved to the edge of Icharni but remained in the Darampur scrub, what seemed like a staging area for forays into the prime habitat for breeding females. In May 1985, we captured and radio-collared him; his lower-left and lower-right incisors measured 7.6 and 6.2 cm, respectively. When we recaptured him to replace a radio collar in January 1986, his incisors measured 8.3 and 6.5 cm. By March 1987, this male had moved to the fringes of the *S. spontaneum* grasslands in the Pipariya area of Icharni, perhaps the prime grazing area in this part of Chitwan. By the middle of 1987 he had become established on the island and had begun to challenge other males. We captured and remeasured him for the third time in

May 1988, six months after he assumed dominance on Icharni; his incisors measured 8.3 and 8 cm.

The Incisor-Size Hypothesis and Reproductive Success

Phenotype and social organization can strongly influence individual variation in reproductive success among males in polygynous species (Clutton-Brock 1988). Among polygynous breeding males, variation in size of weapons (horns, antlers, teeth, claws), body mass, condition, and fighting experience influences the outcome of competition for females and ultimately reproductive success. The presence of nasal or frontal horns distinguishes the five species of rhinoceros from all other mammals. However, horn number, size, and shape and the relative importance of the horn in intraspecific behavior varies among them (Owen-Smith 1988). Perhaps more important is that the five species also differ by the presence, size, and importance of the procumbent mandibular incisors. The three Asian species use mainly the lower incisors in agonistic encounters and other social interactions, and their horn size is smaller than that of the two African species. Conversely, the African species lack procumbent mandibular incisors and instead fight with massive horns.

The distribution and reproductive schedules of females clearly influence social organization and spacing among males. Sections of Icharni Island are adjacent to agricultural lands, and for three months of the year crops may provide 10 to 40% of the rhinos' intake (Jnawali 1995). However, high density of rhinos is not clearly linked to proximity to croplands. Two other areas of Chitwan, far from the park border and far from agricultural activity, support similarly high densities. The extensive S. spontaneum grasslands, suitable forest cover, wallows, and abundant surface water attract females to Icharni.

We have some evidence that female density also influenced the high turnover of dominant males on Icharni during the study period. Although seven males were dominant for short periods on Icharni, one male (M009) remained dominant for more than three years in an area that includes and extends west from Patch 1. As I described earlier, Patch 1 contained fewer females than did Icharni. It may be that where female densities are lower, male dominance persists longer. Alternatively, length of tenure may be related to condition and level of aggressiveness; this male was the most aggressive male in the study area and was responsible for the deaths of several other males. Future studies may determine whether length of tenure is inversely related to female density.

Males in captivity breed as young as five years and may live to be forty (Reynolds 1960). All males but one known to have bred in the study area were estimated to be older than fifteen. Several males showed dental characteristics associated with old age, such as extensive wear on the occlusal surface of the last molar and attrition of the M1 tooth. These males also exhibited other features associated with older animals, such as extensive wrinkles, worn or broken horns, wide horns, and numerous scars; by 1990 one old male (M002) was nearly blind.

Males with broken incisors could not defend themselves against competing males even if their bodies were of similar size. These once-dominant males, as well as young males, were frequently attacked and sometimes killed during encounters on or near Icharni. Thus, regardless of size or experience of breeding males, breakage in incisors put them at risk and led to changes in the dominance hierarchy. Most important, males with broken incisors did not reestablish themselves on Icharni after being evicted.

Evidence of the importance of incisors in social interactions can be drawn from translocation experiments involving rhinoceros relocated from Chitwan to Royal Bardia National Park. A translocated adult male with one broken incisor was able to court and breed females in the absence of males with fully developed intact incisors. Before translocation, this male was subordinate to others and lived in an area of Chitwan where breeding females were scarce. Future translocation efforts provide a natural experiment to test the incisor-size hypothesis by correlating reproductive activity of translocated males with incisor size.

Other factors such as diet, body condition, and fighting experience may influence establishment and maintenance of dominance. Diet is related to dominance because rhinoceros are primarily grazers, and breeding males typically feed in lush *S. spontaneum* grasslands, which provide fresh shoots throughout the year. In contrast, subordinate males live in scrub or more degraded riverine forest that provide less nutritious forage. Habitat selection by weaker males probably is the result of being unable to defend themselves against stronger males (i.e., those with large incisors) that occupy the prime grazing areas.

Differences in fighting experience may be important in separating weak males from strong males. However, younger adult males with large incisors and less experience will defeat males with broken incisors and greater fighting experience. Other factors may impinge on competition for dominance, but maintenance of dominance relates most directly to the condition of the incisors used in fighting.

The development of genetic markers for this species will permit a more rigorous evaluation of the incisor-size hypothesis by providing a more accurate assessment of reproductive success. I look forward to the not-so-distant future when this approach is applied more widely to determine reproductive success among males and females of various age, diet, body condition, and dominance.

Dominance, Tenure, and Lifetime Reproductive Success

This study revealed two important aspects of breeding biology of male rhinoceros that have a direct bearing on estimation of lifetime reproductive success. First, breeding activity is socially delayed among younger adults because older, more powerful males exclude them from grazing areas where breeding females live. Second, the period during which a male assumes dominance, expressed as a fraction of longevity, is short for a long-lived pachyderm. Records of longevity in free-ranging male rhinoceros are lacking, but in the two African species free-ranging males are thought to commonly reach thirty years of age. Thus the long period of adolescence, rapid turnover among dominant males, and the long interbirth interval results in relatively few breeding opportunities while a given male is dominant on Icharni. Within the Chitwan population, scarce breeding opportunities for males may also be a function of population size. With fewer than 150 adult females in the population, once-dominant males forced to live in marginal habitats seldom encounter adult females, which typically remain in or close to floodplain grasslands.

If males with broken incisors are evicted from the study area after only a few months, they have few chances to mate with females. Those that maintained dominance for the longest periods had considerably more opportunities, and in the case of one male (M008), this almost verged on territoriality. The degree and duration for which a male is capable of excluding others may be a reflection of the disparity in incisor size, fighting experience, body mass, and condition between the dominant male and other males. Age structure of the local population of males may also be key: if a stronger male is surrounded by weaker younger males, he may demonstrate territorial behavior. Social organization in male tigers in Chitwan has shown similar variability in duration of tenure and territory size, although this is thought to be largely the result of the strength of the dominant male (Sunquist 1981; David Smith and Charles MacDougal, personal communication, 1988).

Scarce breeding opportunities may partly explain several important features of the biology of the Chitwan population: intense intermale fighting, violent courtship, and high genetic variability. Without accurately determining estrous periods for females during the four-year period, correlating peaks of intermale fighting with estrus is impossible. However, the frequency of fighting that both Laurie (1978) and I observed, and long interbirth intervals, reveal that fights for dominance occur even without the presence of an estrous female in the area. However, periods of estrus likely heighten the intensity of intermale fights.

A puzzling aspect about rhinoceros behavior is why courtship is among the most violent among mammals. Males aggressively pursue females over long distances (2 km) and attack females with their incisors or ram into them in an attempt to subdue them. On one and perhaps two occasions during the study, females died from wounds suffered in attacks by males. Reports of females dying from internal injuries sustained during courtship have also been recorded on at least two occasions in captivity (Michael Dee, personal communication, 1988).

Such behavior seems maladaptive in light of the few breeding opportunities available to males. Why should males risk seriously injuring perhaps the only female that could be receptive at any given time? This remains a mysterious aspect of the behavior of this species. Violent courtship is typical of the Perissodactyla, and other species of rhinoceros engage in courtship chases. Females testing the strength of males may intensify the aggressive nature of these chases by running across home ranges of competing males. Alternatively, courtship chases may be required to trigger sex hormones, enabling males to increase the volume of seminal fluid before ejaculation (Ulysses Seal, personal communication, 1988).

The high turnover among males has important consequences for the maintenance of high genetic variability. Overall, heterozygosity in Chitwan was 9.9%, approaching the highest recorded for free-ranging mammals. Accumulation of genetic variability is commonly attributed to the antiquity and vagility of the species, maintenance of large effective population sizes before a recent population bottleneck, and long generation time (13 years) (chapter 5). Yet it is obvious that high rate of turnover among males on Icharni diminishes reproductive success among individual males. If such high turnover is typical for the remainder of the reserve, it would promote maintenance of the high genetic variability observed in this population.

CHAPTER 8

ENDANGERED PHENOMENA:
RHINOCEROS AS LANDSCAPE ARCHITECTS

Most of the surviving species of big tropical mammals will soon
disappear outside of protected areas. . . . These larger animals . . .
often determine the physical structure and spatial distribution of other
species in ecological communities. When the large animals disappear, the
ecological changes are often swift and profound.
—Michael E. Soulé, "The End of Evolution?"

WHEN I FIRST ARRIVED IN Chitwan, I wondered how any
herbivore could make a dent in the mass of green vegetation created by monsoon floods and rains. One day I asked an elephant
driver why a common tree species, *Trewia nudiflora* (known locally
as *bhellur*), grows in copses, like an archipelago of trees in the midst
of a tall grassland (figure 8.1). "Oh, it's the work of *gaida*," he said,
using the Nepali term for greater one-horned rhinoceros and ges-
turing toward the tree islands in the sea of grass. "Those are old
rhino latrines." Rhinoceros, like many other ungulates and carnivores, return repeatedly to the same place to defecate.

This was a revelation to me. My Western-oriented training in
plant–animal interactions implies a world that has already lost its
terrestrial megaherbivores. Bison, mastodons, giant ground sloths,
North American rhinos, and their allies no longer play their traditional roles as landscape architects. Western plant-ecology texts
stress the importance of local soil type, soil moisture, temperature,
nutrients, and fire as forces that determine the dominant vegetation type. Of course, these factors are also relevant in Chitwan.

FIGURE 8.1. Greater one-horned rhinoceros grazing in front of a copse of *Trewia nudiflora* trees on the floodplain of Icharni Island. The trees are growing on old, and still active, latrines. (Photo by Eric Dinerstein)

However, the tree species that eventually form the canopy of Chitwan's forests are those whose saplings are unpalatable to giant browsers and can withstand the trampling of a 2,000-kg rhino or an elephant's wrecking-ball trunk. Megaherbivores also influence the canopy by ingesting odd plant species that produce large fruits and then dispersing their seeds.

Since the early 1980s, field studies have started to document the changes that occur in the structure and composition of vegetation when giant herbivore populations are below or at the carrying capacity of the habitat. Not surprisingly, a vast body of literature already shows the dramatic effects of large mammalian herbivores on vegetation structure and composition when herbivore populations explode. Numbers of white-tailed deer in the eastern United States have skyrocketed with the extirpation of wolves and mountain lions through most of their range. Overabundant deer populations are responsible for removal of herbaceous plant species (Alverson, Waller, and Solheim 1988), selective browsing of tree seedlings in western forests of the United States (Hanley and Taber 1980), and reduction in ground cover required by nesting migratory songbirds in the eastern United States (McShea and Rappole 2000). Too many elephants in too small an area lead to conversion of African woodlands to savannas and grasslands (Owen-Smith 1988). Overabundant herds of horses can destroy rangelands, burros can devastate desert vegetation, and the list goes on.

My elephant driver's remark about latrines triggered a four-year research program on rhinoceros–plant interactions to clarify several research questions: In a place like Chitwan, what role do giant *native* herbivores play as landscape architects when they exist at levels below carrying capacity? When populations of endangered large mammals decline dramatically, at what point do they become "ecologically extinct"? In other words, is there a population threshold below which endangered species may still persist in low numbers but no longer play their normal ecological role — as consumers of plants, dispersers of seeds, or regulators of other species populations — in the ecosystem to which they belong? I tried to untangle these relationships by asking three specific questions: Do rhinoceros eat fruits, such as *Trewia nudiflora,* that are designed to attract giant herbivores? (This is what Janzen and Martin [1982] dubbed the megafaunal fruit syndrome.*) Does selective browsing by rhinoceros determine which species reach the canopy of riverine forests? How does the virtual extirpation of elephants from the protected area affect vegetation composition and structure?

Contemporary plant ecologists need to consider the extent to which the plant traits that they study today result from selection based on interaction with species from the Pleistocene that are now extinct. This is a new perspective from which to view Chitwan and other reserves that still maintain high densities of native megaherbivores. Aside from their role as havens for endangered species, they serve as refuges for endangered phenomena such as megaherbivore-mediated seed dispersal and the role that giant mammals play as landscape architects.

The Megafauna Fruit Syndrome

Janzen and Martin (1982) proposed that the now-extinct neotropical megafauna once played a major role in the dispersal of woody flora. They argue that the long coexistence of neotropical plants and large frugivores molded the evolution of fruit and seed traits of some plants for consump-

---

*On the ingestion of small-seeded plants by large herbivores, where the foliage serves as the "bait" to attract dispersal agents, see Dinerstein (1989) and Janzen (1984).

tion and dispersal by large mammals. The demise of the gomphotheres, pre-historic horses, proboscideans, ground sloths, and other large ungulates in the Western Hemisphere prohibits a direct test of hypotheses related to seed dispersal, so Janzen (1981a, 1981b, 1982a) used domestic horses and cattle as surrogate Pleistocene frugivores. Experiments with these animals and ob-servations of a captive native tapir support the evolutionary importance of large frugivores in the demography and microdistribution of neotropical plants (Janzen 1981c, 1982b). Howe (1985) has challenged this interpreta-tion as an untestable hypothesis. Had Janzen gone east to Asia, he would have been able to test this hypothesis on a more intact assemblage of mega-herbivores. Perhaps one reason that Janzen stayed in the neotropics is that essential data on frugivory and seed dispersal by large mammals in Africa and Asia are mostly lacking.

The most numerically abundant tree on the floodplain of the Rapti River in east and central Chitwan is *Trewia nudiflora* (Euphorbiaceae) (fig-ure 8.2; table 8.1). During the monsoon season (June–October), *Trewia* fruit ripens in abundance in Chitwan's riverine forests. Bats, birds, and monkeys ignore ripe fruits on the tree, presumably because of their alkaloid content, which makes them bitter; however, rhinoceros devour them with gusto. Deer and domestic cattle eat these fruits to a lesser extent when they fall to the ground. The frequent occurrence of a dense layer of *Trewia* seedlings on rhino latrines in grasslands supports my elephant driver's sug-gestion that rhinos are the major dispersal agents of *Trewia* in Chitwan and an important vector for the establishment of this species on the floodplain.

I decided to investigate the full gamut of the rhinoceros–*Trewia* inter-action. More specifically, I considered the natural history of the plant species, ingestion rates of fruits, gut passage rates of *Trewia* seeds through rhinoceros, seed survival through the gut, effects of seed swallowing and manuring on seedling vigor, rhinoceros-generated seed shadows, the fate of uningested seeds and fruits, and the factors controlling the establishment of *Trewia* on the floodplain. I also examined the distribution of *Trewia* in the absence of rhinoceros and in areas where domestic livestock occur. Both Janzen and his critics consider all these factors to be critical elements in de-tecting the strength of the putative megafaunal fruit syndrome.

The Natural History of *Trewia* Fruits

Large fruit size and a hard endocarp (the fruit tissue encasing the seeds) dis-tinguish *Trewia* from nearly all other fleshy fruits available in Chitwan

A

FIGURE 8.2 *Trewia nudiflora* is the most abundant tree in the riverine forests. (A) Trees are dioecious, with females producing large fruit crops during the monsoon season, and form a low canopy (the author is standing atop the elephant to provide scale). (Courtesy Lori Price)

FIGURE CONTINUED

B

C

FIGURE 8.2 CONTINUED (B) Fruits are large, hard, and green upon ripening, and they contain three to five seeds encased in a pericarp with the texture of a green potato. (C) A captive rhino from Chitwan eats *Trewia* fruits at the Kathmandu zoo as part of an experiment to determine gut passage rates, seed survival through the gut, and effects of gut transit on germination rates. Rubbing has worn away its horn. (Photos by Eric Dinerstein)

TABLE 8.1.   CHARACTERISTICS OF *TREWIA NUDIFLORA* FRUITS

	Mean	SD	N
Wet mass (g)	26.0	4.6	75
Diameter (mm)	37.8	9.3	75
Number of seeds/fruits	3.2	0.9	75
Mass of seeds (g) per fruit	0.4	0.1	20
Seed diameter (mm)	7.8	0.7	14
Water content of pulp (%)	83.8	0.5	8
Pulp/seed ratio (wet mass)	8.4	1.4	8

(Dinerstein 1991b; Dinerstein and Wemmer 1988). Ripe fruits are yellowish green to brown and have the texture of a green potato. Delicious by rhino standards, they are extremely bitter to our taste. The force required to crush *Trewia* seeds is not great, but the hard endocarp of the entire fruit resists cracking (Dinerstein and Wemmer 1988). The force exerted by a 77-kg human stepping on a fruit on a hard surface could not split it open, but many fruits split when stepped on by elephant and rhinoceros.

Trewia is the most common tree in riverine forest, accounting for 48.5% of all stems greater than 3 cm diameter at breast height (dbh) on a 1-ha plot (65 stems/ha) (Dinerstein 1992). *Trewia* is dioecious (male and female flowers born on separate trees). Randomly located plots totaling a 1-ha sample in riverine forest yielded 118 females and 96 males, as well as 88 trees at 4 to 10 cm dbh that did not flower during the census and could not be sexed. Diameter at breast height of reproductive females ranged from 6 to 57.2 cm ($x \pm$ SD = 24.5 $\pm$ 11.3 cm) and from 4 to 65.2 cm in males (21.2 $\pm$ 10.6 cm). Average fruit crop size was 70.2 fruits per tree ($N = 50$; range = 2–1,615 fruits). Fruits begin to ripen during the early monsoon season (June), and peak availability occurs in the mid- to late monsoon season (July–August) (figure 8.3). Ripe fruits undergo a slight color change and fall to the ground. By late August, the high density of *Trewia* trees ensures that fruit carpets the forest floor in the study area.

The Natural History of the Rhinoceros–*Trewia* Interaction

Greater one-horned rhinos relish *Trewia* fruits and actively seek them out during the monsoon season, when the species produces ripe fruit. Twenty-four-hour activity watches ($N = 26$) on habituated rhinos revealed that

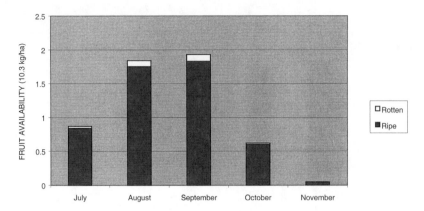

FIGURE 8.3 Ripening of *Trewia* fruits during the monsoon season.

daily consumption of *Trewia* peaked in midmorning just after rhinos emerged from a two- to three-hour wallowing period. Correspondingly, fruit intake was high beneath trees bordering forest wallows. The total number of fruits ingested was accurately determined for five five-minute sampling periods; the average intake ($\pm$ SD) was 17.6 $\pm$ 3.58 fruits per period (range = 14–23). Assuming that 17.6 fruits per five-minute period was representative of the rate of fruit intake, individual rhinoceros ingested 197 fruits (a wet mass of 5.1 kg) every twenty-four hours during the peak fruiting period. The mean number of *Trewia* seeds per kilogram of dung collected monthly from twenty randomly sampled latrines provided an index of *Trewia* consumption. The mean number of intact seeds per kilogram of dung increased in July and August, peaked in September, and declined thereafter (figure 8.4).

A rhino eating the fruits of *Trewia* is merely an example of the natural history of fruit removal, not seed dispersal—a subtle but important difference that the literature on animal–plant interactions often overlooks. What eventually happened to those ingested fruits and seeds? Specifically, how long do seeds stay within the gut once ingested, and how many survive the trip past the large crushing molars and the giant cecum? To answer this question we needed a cooperative, wild-caught rhino in a captive environment. We found our "volunteer," an adult female, at the Kathmandu zoo. Every morning in Chitwan we collected fruits to feed to this rhinoceros and transported them the same day or the following day. Rhinos regularly eat

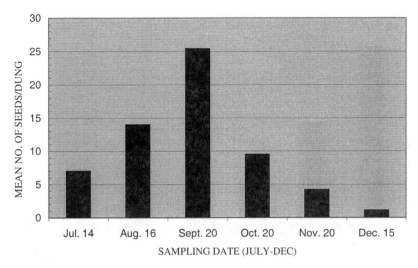

FIGURE 8.4 Seasonal proliferation of intact *Trewia* seeds in rhinoceros dung.

fallen fruit that is at least several weeks old, so I assumed that a one-day delay in feeding had no effect on the experiment. In the first feeding trial, zookeepers placed fruits on a ledge where the captive animal could select or reject them. The rhino relished the fruit and ingested all 114 in 10 minutes. Seeds were first present in the dung 46 hours after ingestion, peak passage occurred between 64 and 88 hours, and the rhino passed the last intact seeds 172 hours (7 days and 4 hours) after ingestion (figure 8.5). This rhino ingested 300 fruits in 51 minutes during the second trial (again the entire amount presented). Seed passage rates for the two feeding experiments were significantly different (see the statistical notes at the end of this chapter). By multiplying the mean number of seeds per fruit (3.2) by the number of fruits fed to the rhinoceros, I estimated that the rhino ingested ≈365 seeds in the first trial and ˜960 seeds in the second. Assuming that we found all intact seeds that the rhino passed, seed mortality was 26.7% and 47.7%, respectively, for the first and second trials. Seed coat fragments also appeared in the dung.

I found that gut treatment, manuring of seeds in dung, and the interaction of the two factors did not have a significant effect on germination of *Trewia* seeds. I included gut passage as a treatment effect along with manuring of seeds because it seemed plausible that if ingestion hastened ger-

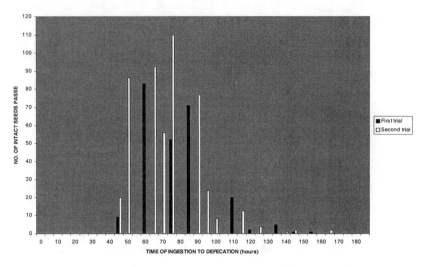

FIGURE 8.5 Transit times for *Trewia* seeds passed by a captive rhinoceros.

mination, young plants grown from passed intact seeds would have a head start on uningested seeds. Manuring of seeds in dung did have a significant positive effect on above-ground dry mass and on dry leaf mass but not on dry root mass. Taproots from seedlings grown in dung were significantly shorter than those from seedlings germinated in soil. Root tissues for seedlings grown in dung remained concentrated in the dung layer, whereas plants grown in soil sent down longer taproots. Gut treatment had no effect on above-ground dry mass of seedlings, dry leaf mass, taproot length, and dry root mass. I observed no significant interaction effects on plant growth of gut treatment and manuring of seeds.

What happens to the seeds if rhinoceros fail to ingest fruits? Sunlight and trampling by heavy mammals play a role here. My experiments show that entire fruits placed in the sunlight rotted in significantly greater numbers than the same number of fruits placed on the forest floor. Placing fruits in the sun also had a significant positive effect on germination and seedling height. The crushing by pachyderms (domesticated elephants) had a significantly positive effect on germination number and seedling height. Crushing the fruit removes the seeds from the hard endocarp. However, given the soft soil during the monsoon, it is unlikely that many ripe fruits on the forest floor are crushed; more likely they simply rot. These observations convinced me that if rhinoceros or some other large herbivore fails to ingest the fruits and, through digestion, remove the seeds from the hard endocarp, recruitment is virtually nil. Fruits left under the parent tree are highly unlikely to yield seeds that germinate in time to grow during the important establishment window created by the monsoon.

If a rhinoceros does ingest the fruits, does it matter where it defecates the seeds? Absolutely. Seedlings on natural *grassland* latrines (figure 8.6) were far more abundant and grew more vigorously than seedlings on natural *forest* latrines. We found a combined total of only 54 seedlings at thirty forest latrines at the end of the growing season, whereas 529 survived at thirty grassland latrines. Of the 54 seedlings at the forest latrines, only 11 had exceeded the cotyledonous leaf stage (an early stage of development) as opposed to 227 for the grassland latrine plants.

Exposure to sunlight is the underlying factor in seedling establishment. I created experimental latrines by arranging rhino dung in discrete miniature latrines. After two months, a significantly greater number of seedlings had survived the monsoon season in sunny than in shady plots. In the latter, most of the seedlings died within four weeks after emergence. Seedlings grown in sun were also significantly larger than those grown in shade. Of

FIGURE 8.6 *Trewia* seedlings and saplings dominate plant cover on rhinoceros latrines in grasslands. Greater one-horned rhinoceros disperse at least thirty-eight species of plants at latrine sites. (Photo by Eric Dinerstein)

TABLE 8.2.   AGE AND HEIGHT OF *TREWIA NUDIFLORA* SEEDLINGS AT THE
END OF THE GROWING SEASON UNDER THE RIVERINE FOREST CANOPY OVER
THREE YEARS

	Density	
Seedling age/height	Number	Number/hectare
First year <16.0 cm tall	429	572.0
First year >16.0 cm tall	126	168.0
Second year	68	90.7
Third year	2	2.7

NOTE: Area sampled = 0.75 ha.

the 600 seedlings planted in the experimental forest plots, only 3 (12.5%) of the 24 seedlings that survived exceeded 16 cm, whereas 498 (85.4%) of the 1,037 seedlings planted in grasslands exceeded 16 cm at the end of the monsoon season. Similarly, seedlings on two-year-old experimental grassland latrines occurred in higher numbers and grew significantly taller than on forest latrines. Most striking is that 19% of the 1,037 seedlings that survived on grassland latrines exceeded 50 cm, and 3% exceeded 2 m in height.* There is further evidence that *Trewia* seedlings are shade intolerant. A census of a 0.75-ha patch of riverine forest indicated that recruitment of *Trewia* seedlings under the canopy was remarkably low (table 8.2). The entire 0.75-ha area, sampled two months after the most recent monsoon, yielded 429 seedlings shorter than 16 cm, 126 seedlings taller than 16 cm, only 68 second-year seedlings, and 2 third-year seedlings. Sapling and tree diameters, ranked by size classes, confirm low regeneration (figure 8.7). Of 328 individuals encountered in the twenty plots, we found only 10 (3%) saplings that were taller than 50 cm.

Trewia is not the only species that takes advantage of latrines as a site of establishment. The annual legume *Cassia tora* (Leguminaceae), an abundant weed in overgrazed pastures and scrub outside the park, contributed 32.4% of cover. *Trewia* seedlings contributed 24.1%, whereas saplings and trees covered 11.8%. On latrines farther than 3 km from croplands, *Cassia* and other annuals were rare to absent, and *Trewia* seedlings dominated. Overall, we encountered thirty-eight species, including four trees, five grasses, sixteen shrubs, six herbaceous plants, and seven herbaceous climbers, at latrine sites (Dinerstein 1991b). Other vertebrates also occur in these latrines, including the endangered turtle *Melanochelys* (Dinerstein, Zug, and Mitchell 1987).

Unfortunately for *Trewia*, during its fruiting season rhinoceros tend to defecate in the shade at forest latrines rather than out in the grassland latrines. Thirty percent of all observed defecations occurred in grasslands, and 56% occurred in riverine forest during the *Trewia* fruiting season ($N = 23$ defecations observed during 24-hour watches). At least when a rhinoceros defecates at a latrine, the addition of rich fertilizer to young *Trewia* plants is prodigious: mean ($\pm$ SD) dung mass per bout during *Trewia*'s seasonal fruit fall was 11.1 ± 6.7 kg (range = 3.7–24.3 kg; $N = 9$). Latrines typically range in width from 3 to 10 m for some of the largest and most active sites.

Soon after fruit fall and manuring of seeds in latrines, *Macroceroea grandis* and *Iphita limbata* (Largidae), two abundant species of hemipteran seed predators, attack *Trewia* seeds (figure 8.8). Adults of both species are strong fliers; these bugs quickly find experimental grassland latrines deposited in tall grassland as far as 300 m beyond the forest edge. Adults and juveniles were frequently observed climbing into rotting fruits to attack the seeds. Domestic cattle were the only mammalian herbivores of *Trewia* seedlings.

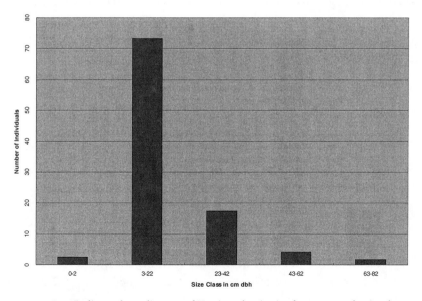

FIGURE 8.7 Sapling and tree diameter of *Trewia* under riverine forest canopy, by size class; dbh, diameter at breast height.

FIGURE 8.8 *Macroceroea grandis*, pictured here, and *Iphita limbata* (Largidae) commonly attack *Trewia* seeds. (Photo by Eric Dinerstein)

In grassland latrines, most seeds that escaped seed predators germinated. The seed load deposited in one rhinoceros latrine exceeded 5,600 intact seeds, and initial seedling density exceeded 15 seedlings per 100 cm². A small percentage of seedlings died from trampling when dominant males displayed at latrine sites, by defecating and dragging their hooves through the dung. The high density of survivors suggests intense intraspecific competition among seedlings for light and nutrients in the rich dung.

What, then, is the magnitude of the rhinoceros–*Trewia* interaction? With about sixty rhinoceros in a 10.5-km² mix of floodplain and riverine forest, I estimated that the study population may have ingested 306 kg of fruit per day, and by the end of the monsoon season the rhinoceros had manured at least 1.5×10^6 seeds into grassland latrines. A single rhino, with *Trewia* composing only 10% of its ingesta, can still be carrying ≈10 kg of seeds and fruit tissues in its gut (probably more than 1,000 seeds/day). To put the scale of this interaction into perspective, the seed load found in 10% of the ingesta of a single rhino roughly equals the ingesta held by a colony of about 2,000 Jamaican fruit bats (*Artibeus jamaicensis*), perhaps the best-studied and most ubiquitous neotropical mammalian frugivore. Furthermore, few contemporary frugivores can match rhinos in the quantity of seed-rich manure (up to 24.3 kg) produced in a single defecation. The heavy manure loads are significant. My experiments revealed that seeds defecated into latrines received a substantial boost from the manure in which they germinate and that seeds defecated in grassland latrines can grow into robust saplings after only two monsoon seasons.

So what prevents riverine grasslands from developing rapidly into *Trewia*-dominated woodlands and ultimately *Trewia*-dominated riverine forests? Heavy floods during the monsoon and fire during the hot season emerged as the two most significant factors that altered the *Trewia* seed shadow created by rhinoceros in grasslands. Five hundred marked seedlings that were scattered among fifteen grassland latrines all died after a major flood buried them in a 50-cm layer of silt. However, floods also washed *Trewia* seeds and fruits out of forest latrines and from the forest floor and onto the floodplain.

Fires from lightning strikes during the late hot season were probably frequent in prehistoric times and still occur. However, most fires within the park are now caused by local grass cutters and by elephant drivers who set them so that they can more easily harvest the canes of elephants grasses and improve grazing and game viewing, respectively. These fires, set annually between January and May, limit recruitment on grassland latrines. The experiments with 500 marked seedlings at fifteen grassland latrines revealed that mortality is 100% if a hot fire burns across a latrine. Saplings that were protected from fire for two growing seasons exceeded 2 m in height. At this level of development, they survived grass fires under two conditions: when fires occurred late in the afternoon when the evening dew would extinguish the grass fires, or when fuel, in the form of dead grass around the saplings, was low. It is likely that naturally occurring fires in prehistoric times burned smaller areas of grasslands, resulting in a higher rate of *Trewia* recruitment on grassland latrines.

Monsoon floods do not inundate all grasslands, nor do all areas of grasslands burn each year. My observations suggest that if seedlings in grassland latrines can escape the effects of floods and fire for one to two years, they are well on their way to colonizing the floodplain. Aerial photographs of the study area in 1975 compared with recent surveys depict rapid colonization of *Trewia* on Chitwan's floodplain. Verbal accounts of earlier researchers (W. A. Laurie and David Smith, personal communications, 1988) and project elephant drivers corroborate the swift establishment of *Trewia*, particularly on rhino latrines in grasslands. Fifteen years before our survey, much of the current study area was a major river course. When the river shifted, grasses colonized the exposed silt, and by 1986 small groves of

Trewia dotted the floodplain; many of these groves contain active rhino latrines. In 1970 one of these areas, known locally as Bhellurghari (*Trewia*
area or site in the Tharu language), was a collection of rhinoceros latrines
on the edge of a river terrace with many small seedlings and saplings emerging from the dung piles. A 1987 survey of *Trewia* found 169 stems greater
than 3 cm dbh. Mean (± SD) dbh for *Trewia* in this area was 33.4 ± 20.8
cm, and mean (± SD) height was 13.2 ± 6.5 m (range = 2.5–29.2 m), further evidence that *Trewia* is one of the fastest-growing tree species on the
floodplain.

Despite the magnitude of the rhinoceros–*Trewia* interaction in Chitwan, frugivory remains nothing more than an interesting sideline of rhinoceros natural history unless it can be demonstrated that ingestion of
Trewia by large mammals increases plant fitness. In sum, the data demonstrate that from ingestion to seedling establishment on the floodplain, large
mammals influence the demography and microgeography of the contemporary *Trewia* population. First, only large mammals disperse *Trewia*. Second, *Trewia* seeds are packed into a hard endocarp, so seeds that are not
freed from the fruit tissue through digestion do not germinate before the
end of the monsoon season, the optimal germination and growing period.
Third, despite the high density of *Trewia* in riverine forest and the enormous fruit fall during the monsoon season, recruitment under the canopy
is poor.

Assessing the influence of rhinoceros on the demography and distribution of *Trewia* requires a comparison of similar grassland habitats with and
without rhinos. Floodplain dynamics are rapid throughout the Terai (Dinerstein 1987); thus if other large wild herbivores effectively disperse *Trewia*,
it should be a common part of the landscape. For the most part, this is so.
A reserve in Bihar, India, has no rhinos, but *Trewia* flourishes through dispersion by Indian bison. In Jaldhapara Sanctuary, in West Bengal, where
only a few rhinoceros remain, domestic buffalo disperse *Trewia*. In another
part of the Nepalese Terai and in northern India, *Trewia* is a common tree,
but in the absence of wild herbivores domestic cows disseminate *Trewia*
fruits.

Resident populations of rhinoceros disappeared from what is now
Royal Bardia National Park in western Nepal more than 150 years ago, and
Trewia is a rare to uncommon tree restricted to a few watercourses. We observed no germination in grasslands away from riverbanks even though axis
deer, known *Trewia* eaters, are abundant in Bardia. Since the reintroduction

of thirteen rhinoceros to Bardia in 1986 (Mishra and Dinerstein 1987), preliminary data suggest that the recruitment of *Trewia* on grassland latrines has begun (Gagar Singh, personal communication, 1988). More than a decade later, greater recruitment of *Trewia* in Bardia should be evident, assuming that extensive grassland fires do not retard seedling establishment.

Rhinoceros figure prominently in *Trewia* dispersal and most probably in the evolution of fruit traits. Yet, because other animals facilitate dispersal, it is unrealistic to expect that *Trewia* would disappear from the floodplain in the absence of rhinos. The absence of tight dependence of a given fruit on dispersal by a given frugivore may be attributed to (1) the advantages for the plant to appeal to a wide spectrum of animals, (2) similar nutritional requirements among the fruit eaters, (3) opportunistic feeding by frugivores in search of an easy (undefended) meal, (4) the difficulty of evolving cues to limit detection and palatability for nontarget frugivores versus target species, and (5) the loss of large frugivores over ecological and evolutionary time (Howe 1984; Thompson 1982; Wheelwright and Orians 1982). Nonetheless, megafauna–fruit interactions are common in certain natural forests: Alexandré (1978), working in the Taï Forest of the Ivory Coast, listed thirty species of trees and shrubs for which the African elephant seemed to be the only dispersal agent. Other such interactions between large indehiscent fruits and large terrestrial mammals probably await discovery in Southeast Asian tropical forests wherever such mammals are still common.

The most likely explanation for the occurrence of large indehiscent fruits, like *Trewia*, in Earth's woody flora is that these fruits evolved in response to megaherbivores. In Chitwan woody plants producing such fruits constitute a relatively small proportion, probably less than 10%, of the local flora (Dinerstein 1991b). Nevertheless, the short species list, from a taxonomist's perspective, does not obscure its significance to the ecologist.

Selective Browsing by Megaherbivores and the Effects on Forest Structure

Several large Asian herbivores feed extensively on foliage and stems (Laurie 1978; Olivier 1978), significantly distorting tree growth (Mueller-Dombois 1972). However, ecologists have paid little attention to the effect of browsing behavior on forest structure. The contemporary guild of large browsers includes Asiatic elephant, greater one-horned rhinoceros, Javan rhinoceros, Sumatran rhinoceros, Indian bison, and banteng. These mammals have co-existed with forest plants for millennia, and it is reasonable to assume that chronic herbivory has been an important selective force on certain Asian plant species, as it has for plants in the neotropics (Janzen 1986) and in Africa (Owen-Smith 1987).

Another effect of giant mammals on forest structure became clear to me during a November evening in my first winter in Chitwan. Each night rhinoceros would walk through the riverine forest bordering our camp and push down saplings of the tree *Litsea monopetala* Roxb. (Lauraceae) (known locally as *kutmiro*). The whoosh and snap of large saplings being bent to the ground was followed by the sound of massive molars grinding leaves and stems. By the following month, the understory looked like a war zone, with mangled and twisted stumps of *Litsea* scattered throughout the forest (figure 8.9).

I decided to take a detailed look at a rhino–plant interaction to further elucidate the rhino's influence on forest structure and canopy composition. The question became whether chronic browsing and bending of *Litsea* by rhinos prevented most *Litsea* individuals from reaching the canopy. To an-

FIGURE 8.9 A *Litsea monopetala* sapling browsed and trampled by *Rhinoceros unicornis* in Royal Chitwan National Park. (Photo by Eric Dinerstein)

swer this question, I compared growth response of *Litsea* saplings within and outside protective exclosures.

The Natural History of the Rhinoceros–*Litsea* Interaction

Litsea is a canopy species reaching 25 m in riverine forest. In certain stands, *Litsea* saplings occur in high densities and are the most common species of sapling we encountered. Rhinoceros often trample to the ground *Litsea* stems smaller than 15 cm dbh. When foraging on *Litsea*, adult rhinoceros can either remove the leaves and stems (up to 100 g wet mass) with each bite or pluck individual leaves with the upper prehensile lip. These trees are evergreen, with moderate leaf fall beginning in November (Dinerstein 1987). *Litsea* produces new leaves in mid-February, coinciding with the period when new grass shoots emerge on the adjacent floodplain. Rhinoceros largely abandon browsing on *Litsea* and return to the grassland at this time.

About sixty rhinoceros used the 7-km^2 study area in which we inventoried *Litsea* stems (appendix A). Data from movements of radio-collared individuals and from population censuses revealed that from November until February, one to six individuals per night fed on *Litsea* in the forest tract containing the exclosures. Exclosure studies revealed that browsing and trampling by rhinoceros had a significant negative effect on sapling growth (table 8.3). Exclosures kept out rhinoceros but were open to smaller herbivores, none of which browse or trample *Litsea*. After three years of study, the grand mean ± 1 SD for sapling height for all protected plots was 4.89 $\pm$ 0.28 m; thirty-one protected saplings were 6 to 7 m high, eleven were 7 to 8 m, and four exceeded 8 m. In contrast, after three years the grand mean

TABLE 8.3.   NESTED ONE-WAY ANALYSIS OF VARIANCE OF SAPLING HEIGHT OF *LITSEA MONOPETALA*: COMPARISON OF UNPROTECTED SAPLINGS IN EXCLOSURES FOR THREE YEARS, ROYAL CHITWAN NATIONAL PARK

Source	SS	d.f.	MS	F	P
Total	521.9	179	2.9	—	—
Protection	211.7	1	211.7	105.9	.00001
Replicate	20.0	10	2.0	1.2	.3235
Error	290.3	168	1.7	—	—

NOTE: SS, sums of squares; d.f., degrees of freedom; MS, mean sums of squares; F, value of the test statistic; P, probability.

±1 SD for unprotected saplings was 2.79 ± 0.13 m, the tallest plant was 5.5 m, and only six stems were more than 4 m high.

I inventoried all stems of woody plants in ten forest plots of 0.5 ha each. We counted 2,073 woody stems; 96% were saplings, small trees, and canopy individuals of six species. *Litsea* and *Mallotus philippinensis* (another species browsed and trampled by rhinoceros) accounted for 33% of all stems. The size structure of *Litsea* was highly skewed toward trampled saplings; only 30 (5%) of the 574 *Litsea* stems measured were 3.5 m tall or taller. All 544 understory stems showed signs of moderate to heavy browsing and trampling by rhinoceros.

In response to intensive herbivory, saplings of *Litsea* sprout readily in February, producing new shoots along browsed and bent stems and from locations where old stems were snapped. Partially broken and bent trunks often spread horizontally in several directions and may be more than 3 m long (figure 8.9). Browsing and trampling stimulate production of new leaves and stems in the plant's lower reaches (below 2 m). Saplings chewed and pruned by rhinoceros produced significantly more leafy branches below the browse line than did unbrowsed saplings (table 8.4). This is because most of the new growth on protected saplings occurred at the upper edge of the crown and thus was higher than 2 m instead of at the base of the tree. Increase in leaf abundance below 2 m on browsed saplings is related not to phenological changes induced by herbivory but to manipulation of branch height and growth.

TABLE 8.4. NESTED ONE-WAY ANALYSIS OF VARIANCE OF LEAFY BRANCH PRODUCTION OF *LITSEA MONOPETALA*: COMPARISON OF UNPROTECTED SAPLINGS AND SAPLINGS IN EXCLOSURES FOR THREE YEARS, ROYAL CHITWAN NATIONAL PARK

Source	SS	d.f.	MS	F	P
Total	12,195.9	11	1,108.7	—	—
Protection	7,805.1	1	7,805.1	64.47	.0001
Replicate	4,390.9	10	439.1	3.63	.0001
Error	36,806.4	304	121.1	—	—

NOTE: Analysis considered only leafy branch production at heights of less than 2 m. SS, sums of squares; d.f., degrees of freedom; MS, mean sums of squares; F, value of the test statistic; P, probability.

The Effects of Giant Browsers on Vegetation Structure and Composition

Exclosure studies addressing the effects of native and introduced herbivores on seedling survival and vegetation structure are common (Alverson, Waller, and Solheim 1988; Graf and Nichols 1967; Hanley and Taber 1980). However, most studies focus on the effects of selective browsing by small to medium-size ruminants or lagomorphs. Large mass, extended reach, and great strength can intensify the potential effects that foraging rhinoceros and elephants exert on tree growth and architecture.

Exclosure studies in the riverine forests of Chitwan clearly demonstrate that browsing and trampling by rhinos inhibit vertical growth of *Litsea* (figure 8.10). Rhinos' browsing and trampling of *Litsea* and *Mallotus* saplings affect forest structure, in particular because saplings occur in high densities in riverine forest. In forest patches where *Litsea* and *Mallotus* are common, the first stratum of bent or prostrate saplings of these species forms a low canopy. Above this mat of saplings exists a gap occupied by a few adult *Litsea*, *Mallotus*, or *Bombax ceiba*, surrounded by a canopy of *Trewia*, *Ehretia elliptica*, and *Premna* spp. In the absence of rhinos, this space undoubtedly

FIGURE 8.10 *Litsea monopetala* saplings five years after protection from browsing by rhinoceros (exclosure fencing has been removed). Note the man at base of the trees for scale. (Photo by Eric Dinerstein)

would be occupied by a greater number of nearly mature or adult individuals of *Litsea* and *Mallotus*.

The observed intensity of bending and browsing of saplings might appear to indicate that the rhino population is beyond the area's carrying capacity. On the contrary, as of 2001, the Chitwan population still seemed to be below carrying capacity. The exclosure study was conducted in 1986 when rhino numbers were even lower. Even at reduced population levels, the interactions described here between rhinoceros and woody plants clearly suggest a significant evolutionary effect of selective browsing by large mammals with potential cumulative effects on forest structure and canopy composition. The influence of giant browsers may be particularly conspicuous on Southeast Asian floodplains, where tree species diversity is low and large browser biomass in riverine forest–grassland mosaics approaches the highest values calculated to date in protected reserves in Asia (Dinerstein 1980; Eisenberg and Seidensticker 1976). Thus it is clear that successful canopy growth in Chitwan's forests is not simply a function of competition for light, water, and nutrients. Saplings that make it to the canopy must also be unpalatable to giant browsers.

Evidence of Ecological Extinction in Megaherbivores

What happens when population levels of giant herbivores fall so low that they can no longer fulfill their role in maintaining the structure and composition of the native vegetation? Wild elephants in the Terai provide a good illustration. Wild elephants probably had a dramatic effect on Chitwan's landscape, especially in the savannas dominated by *Bombax ceiba* (figure 8.11). In many ways *Bombax* (also known as the kapok tree) plays an ecological role in Asian savannas similar to that of the baobab (*Adansonia digitata,* also a member of the Bombacaceae) in Africa. Elephants chew the inner cambium of *B. ceiba* and can girdle a large tree quickly. Domestic elephants left untended while feeding in these savannas often seek out *Bombax* trees. They are actually fed *Bombax* branches during the dry season when new grass is less available. Some researchers believe that when elephants were more abundant in the early twentieth century, savannas probably had few *Bombax* trees in them (John Seidensticker, personal communication, 1988). An interesting test of this hypothesis will be possible in a few years because wild elephants are increasing in number and spending more time in the part of the park where such savannas are common. Careful monitor-

FIGURE 8.11 A savanna in Royal Chitwan National Park with *Bombax ceiba* trees that are fifty to seventy-five years old. The dominance of tall individuals of this species often is a sign of the absence of wild elephants. (Photo by Eric Dinerstein)

ing of elephant populations in Chitwan will help determine at what densities their role as landscape architects will resume.

Other effects of large mammals on vegetation structure are more speculative but worth mentioning. During the 1980s John Lehmkuhl, who was studying grassland ecology, and I noticed that wild boar dug up the root tissues of the tussock-forming grass *Saccharum spontaneum*. Excavation of the tubers by the boar felled the species, which reaches a height of 3 to 4 m, creating ubiquitous patches of bare soil. Within a growing season, dense stands of largely unpalatable shrubs seldom browsed by rhinoceros or other ungulates (*Artemesia vulgaris, Pogostemon benghalense, Colebrookia oppostifolia,* and *Clerodendron viscosum*) colonized these sites, precluding the regeneration of grasses. We suspected that the result was a reduction of grazing areas for larger ungulates. However, the dramatic flood in the monsoon season of 1993 altered the microdistribution of plants dramatically. The areas covered by these invading unpalatable shrubs were either washed away by the most serious flood observed in twenty-five years or buried in several meters of sand. The following year, *S. spontaneum* recolonized the bare silt layer in parts of the same locations. These events stress the importance of long-term habitat studies that span major disturbances within a system. Short-term field studies that fail to overlap with these infrequent large-scale events may be recording only epiphenomena.

Statistical Notes

Testing the Megafaunal Seed Dispersal Syndrome Hypothesis

1. Seed passage rates of *Trewia* for the two feeding experiments were significantly different (Kolmogorov-Smirnov two-sample test, $p < 0.05$).
2. Gut treatment, manuring of seeds in dung, and the interaction of the two factors did not have a significant effect on germination of *Trewia* seeds (Kruskal-Wallis two-way ANOVA, $N = 3$, $p < 0.05$).
3. Manuring of seeds in dung had a significant positive effect on above-ground dry mass (Kruskal-Wallis ANOVA, $N = 3$, $H = 8.308$, $p < 0.005$).
4. Manuring of seeds in dung had a significant positive effect on dry leaf mass ($H = 8.308$, $p < 0.005$).
5. Manuring of seeds in dung did not have a significant positive effect on dry root mass ($H = 1.442$, $p < 0.10$).
6. Taproots from seedlings grown in dung were significantly shorter than those from seedlings germinated in soil ($H = 8.308$, $p < 0.005$).
7. Entire fruits placed in the sun rotted in significantly greater numbers than the same number of fruits placed on the forest floor (Mann-Whitney U test, $N = 10$, $U = 80$, $p < 0.05$).
8. Placement of *Trewia* fruits in the sun also had significant positive effects on germination (NPAR randomized block ANOVA, $N = 10$, $p < 0.001$).
9. Placement of *Trewia* fruits in the sun also had significant positive effects on seedling height (NPAR randomized block ANOVA, $N = 10$, $p < 0.001$).
10. Removing seeds from the hard endocarp by simulating crushing by pachyderms had a significant positive effect on germination number (nonparametric [NPAR] randomized block ANOVA, $N = 10$, $p < 0.05$; see Zar 1984: 228).

11. Removing seeds from the hard endocarp by simulating crushing by pachyderms had a significant positive effect on seedling height (NPAR randomized block ANOVA, $N = 10$, $p < 0.01$).

12. Seedlings on natural grassland latrines were far more abundant and grew more vigorously than seedlings on forest latrines (Mann-Whitney U tests with normal approximation: $N = 30$, $Z = 5.582$, $p < 0.05$; and $N = 30$, $Z = 3.877$, $p < 0.05$, respectively).

13. On two-month-old experimental latrines, a significantly greater number of seedlings on sun plots survived the monsoon until the end of the growing season than on shade plots, where most of the seedlings died within four weeks after emergence (Mann-Whitney U test, $N = 20$ and 14, $U = 239.5$, $p < 0.001$).

14. Seedlings grown in the sun were also significantly larger than those grown in the shade ($N = 20$ and 14, $U = 280$, $p < 0.001$). Of the 600 seedlings planted in the forest plots, only 3 (12.5%) of the 24 seedlings that survived exceeded 16 cm, whereas 498 (85.4%) of the 1,000 seedlings planted in grasslands exceeded 16 cm.

15. Similarly, seedlings on two-year-old experimental grassland latrines occurred in higher numbers (Mann-Whitney U test, $N = 10$, $U = 100$, $p < 0.001$).

16. Seedlings on two-year-old experimental grassland latrines also grew significantly taller than on forest latrines ($N = 10$, $U = 90$, $p = 0.002$).

PART III

The Recovery of Endangered Large Mammal Populations and Their Habitats in Asia

A FORMER WARDEN IN CHITWAN recalls how angry villagers often confronted him during periodic meetings with local communities. They demanded greater access to the park to satisfy their need for firewood and grazing areas for their livestock. Eventually, they would always ask, "Who is more important, people or rhinoceros?" And he would reply, "There are only four hundred rhinoceros in Chitwan, and there are over eighty thousand human residents. I am sorry, but it is my duty to protect the rights of the minority."

This was a courageous response, but the situation exemplifies the conundrum that exists in all communities that view conservation and development as an either/or proposition. Our inability to introduce effective approaches for achieving both goals is one of the most critical barriers that conservationists face. While we fail to overcome this problem, we will continue to lose our dwindling populations of megafauna and their habitat. Throughout much of Asia and Africa, conservation efforts on behalf of rhinoceros and other giant mammals have become temporary holding actions and, in some cases, retreat. The World Wildlife Fund recently completed an analysis of where Asian rhinoceros and elephant populations are likely to persist into the twenty-second century. The list is a disturbingly small subset of the current distribution of these species (World Wildlife Fund 2002).

In part III, I spotlight the remarkable achievements of a developing nation in large vertebrate conservation. The efforts of many people to conserve greater one-horned rhinos, tigers, and other endangered species in Royal Chitwan National Park and its buffer zone have been truly inspiring. The recovery program evolved from a tree nursery situated on 1 ha of private land in 1987 to the creation in 2001 of the Terai Arc, a government-sanctioned plan for the ecological restoration of an entire ecoregion and its unique large mammal fauna.

The important message here is that a country like Nepal, extremely poor and lacking in infrastructure, is restoring endangered species populations. Other countries, both developed and undeveloped, have no excuse not to try.

Specialists interested in the demography of rhinoceros populations (chapter 5), megafauna–plant interactions (chapter 8), or other more traditional scientific topics might find the transition to a discussion of ecotourism programs somewhat surprising. However, many reserves in South Asia are too small to conserve viable populations of area-sensitive species such as rhinoceros, tigers, and elephants. In South Asia, the only means of redesigning reserves is through landscape-scale conservation: large core areas surrounded by well-managed buffer zones linked by suitable dispersal corridors (Dinerstein et al. 1997). In densely populated areas, such as in much of the Terai, conservation at the landscape scale is possible only by involving local people and offering appropriate economic incentives to gain their active participation in the conservation of large, potentially dangerous, animals. Ecotourism, if properly designed and managed, can be an important incentive.

CHAPTER 9

DOES PRIVATELY OWNED ECOTOURISM
SUPPORT CONSERVATION OF
CHARISMATIC LARGE MAMMALS?

The future of greater one-horned rhinoceros, tigers, and other
megafauna will only be secure when in the eyes of local people they are
worth more alive than dead.
—John Seidensticker, Sarah Christie, and Peter Jackson, eds.,
Riding the Tiger

I WAS SITTING ON THE BACK of my elephant, Mel Kali,
watching a rhino graze a short distance away. Above us, a flock of
black ibis was flying in tight formation as it returned to Chitwan
from the south, the birds' melancholy cries signaling the end of the
monsoon. The rain clouds had dispersed, unveiling the summits
of the towering Himalayan chain that were gleaming in a coat of
fresh snow. Annapurna, Mount Dhaulagiri, Himalchuli — even
the names of these great mountains are inspiring, not to mention
their extraordinary heights. I experienced a profound sense of
exhilaration that October day as I sat and observed the rhinos
framed by the majestic backdrop of the world's highest mountain
peaks. Where else in the world can one find such a snapshot offering so vivid a natural contrast?

The goal of tourism developers is to translate reveries such as
mine into economic profit. And in terms of ecotourism, Nepal is
seemingly a gold mine. Nepal already boasts Mount Everest, one
of the world's most famous landmarks; its base camp has become
a mecca for trekkers. For the wildlife enthusiasts from industrial-

ized countries, Chitwan has become the major destination in Asia. Tourists flock there in the hope of glimpsing a tiger in the wild and to ride elephants in search of rhinos. Tigers are elusive nocturnal creatures and are rarely seen anywhere in the subcontinent. But visitors are virtually guaranteed to have their fill of viewing greater one-horned rhinos in Chitwan and Bardia or in Kaziranga National Park in Assam, India. However, neither Bardia nor Kaziranga is as accessible as Chitwan, which can be reached by a twenty-minute plane trip or four-hour bus ride from Kathmandu. And neither is set up to handle the influx of tourists that they could lure from Chitwan.

Charismatic large mammals have become a powerful magnet for the growing worldwide nature tourism industry. The giant vertebrates are appealing for several reasons. First, the giant mammals native to most industrialized nations have long since disappeared. Second, large tropical mammals still survive in what are considered exotic settings. Third, the sheer size and power of these species inspire awe. Finally, a growing number of conservation-minded tourists have a romantic or spiritual attachment to large wild animals. To see them in the wild, tourists must bump about in four-wheel-drive vehicles in the *Acacia* savannas and miombo woodlands of Africa or board whale-watching boats that ply coastal seas and lagoons of the Sea of Cortez or ride elephants through the tallgrass floodplain of Chitwan. Of course, the rustic transport adds to the romance when one is back home and riding the subway or driving to work.

Unlike agriculture, commercial logging, mining, or oil exploration, ecotourism has the potential to bring large sums of money to remote areas while doing little harm to endangered species and habitats. With a mandate to alleviate poverty, development agencies and foundations, along with many conservationists, view ecotourism as an important component of the conservation toolbox. Ecotourism also logically fits in with experiments designed to provide benefits for local stakeholders and, in theory, make them partners in saving species and wildlands (Biodiversity Conservation Network 1995; Western and Wright 1994). Two economic conditions must be met to ensure successful integration of conservation and development goals. First are economic incentives that provide immediate benefits to local people. Second, project incentives must be substantial enough to offset threats to biodiversity (Bookbinder et al. 1998; Dinerstein et al. 1999). The extent to which these economic incentives are derived from conservation activities, rather than from direct financial compensation (e.g., paying a farmer for damage to crops by rhinoceros and elephants or for domestic cattle killed by tigers), is even more important for sustainability.

The Human-Dominated Landscape

Local residents have largely transformed into rice paddies and other types of cultivation the grasslands and riverine forest mosaics of the Chitwan Valley, which support the highest levels of wild ungulate biomass in Asia. Across most of South Asia, burgeoning human populations require more and more land for food production and livestock. The livestock competes with wild ungulates for forage. This is the case with Chitwan as well.

Thirty-six village development committees, which support a total population of more than 260,000 people, border Chitwan on three sides (National Planning Commission Secretariat 1991). Villages have especially high densities along the northern boundary of the park and its buffer zone. High natural population growth poses a serious threat to long-term conservation. With a natural increase rate of 2.7 (table 9.1), Chitwan's population

TABLE 9.1.   DEMOGRAPHIC STATISTICS FOR CHITWAN, NEPAL, AND SOUTH-CENTRAL ASIA

Demographic data	Chitwan	Nepal	South-Central Asia
Birthrate	33	38	31
Death rate	5	14	10
Percentage of population younger than 15	22	42	38
Natural rate of increase	2.7	2.1	2.4
Total fertility rate	4.5	5.8	3.8
Population doubling time	24	33	29
Population density (km^2)	161	165	130
Infant mortality rate (per 1,000 births/yr.)	190	102	79

NOTE: "Chitwan" refers to the seven village development committees sampled. The statistics for Chitwan were calculated from data collected in a 1994 household survey (Bookbinder et. al. 1998). For Chitwan, the infant mortality rate is a general estimation based on births and deaths over a decade, not only for 1995. Population density for Chitwan is based on 1991 survey conducted by the Chitwan District government. Statistics for Nepal and south-central Asia for 1995 were compiled by the Population Reference Bureau (1995).

will double by 2018. Not surprisingly, given such high growth, the population is also very young; 22% of the population is younger than fifteen, and less than 6% is fifty or older. Twenty-four percent of women living in the area are in their reproductive years (15–45 years), and the average number of children a woman will have is four to five, assuming that current age-specific rates remain constant throughout her childbearing years.

Compounding the effects of the natural growth rate are the environmental effects of squatters and in-migrants. Squatters have no legal place to settle. A relatively strict land-tenure system apportions private land; a household survey showed that approximately 87% (865) of the villagers were landowners and held title (Dinerstein et al. 1999). In the same area the in-migration rate was only 0.17. One hundred thirty villagers in the survey were landless, with the greatest number (32%) concentrated in a single village (Piple). The survey showed that loss of property and farmland from flooding in 1993 was the primary reason for landlessness in this village. The shifting course of the monsoon-swollen rivers of the Terai often claims agricultural land or deposits alluvial sands, making farming impossible. Some landless people end up in resettlement projects. Parts of Chitwan's buffer zone are bordered by areas set aside for flood victims from all over Nepal.

The officially demarcated buffer zone for Chitwan National Park contains 750 km^2 of land. Remarkably, about 60% of the buffer zone remains forested, although some parts are seriously degraded (Arun Rijal, personal communication, 1998). An intensive study designed to examine the effects of ecotourism concentrated on seven villages (the exception was Padampur, which was being resettled) along the northern border of the park, with an estimated combined population of 64,000 (figure 9.1). The majority of villagers are subsistence-level farmers of Tharu descent. The remaining villagers come from hill tribes that resettled in the Terai once malaria was brought under control in the 1950s. Wealthy ecotourists who fly into Tiger Tops or one of the other luxury hotels rarely encounter the widespread poverty in the park's buffer zone. The annual per capita income is approximately $150, and more than half the population earns less than $100 annually (Keiter 1995) (all dollar amounts are expressed in U.S. dollars). Locals still rely on both the buffer zone and, to a lesser extent, the park for firewood and pasture land for livestock. (Chapter 10 discusses human use of the wildlife habitats.) Health care and educational opportunities are slowly improving around Chitwan. However, those who live in the more remote parts must travel at least 20 km to a health post. In more isolated areas only the most rudimentary primary-level education is available.

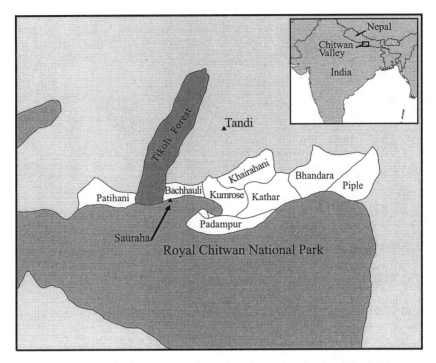

FIGURE 9.1 Village development committees along the northern border of Royal Chitwan National Park, the area studied to determine the effects of ecotourism on local incomes. (Adapted from Bookbinder et al. 1998:1401, fig. L)

Assessment of the Ecotourism Industry

More than 96,000 tourists visited Chitwan in 1997. The tourist boom began in 1981, stimulating a dramatic increase in hotel construction in Sauraha, a small rural ward in the Bachhauli Village Committee on the border of Chitwan. Sauraha is also the epicenter of ecotourism—the main entrance to the park and the highest concentration of ecotourist hotels are located in this ward. Since the early 1980s, the number of privately owned ecotourist hotels has steadily increased. In October 1995, Sauraha had forty-six relatively inexpensive hotels, and seven larger, more expensive hotels were located inside the park (figure 9.2). Three were built before 1980, thirty-one in the 1980s, and nineteen in the 1990s. Hotel development continued as of 1998, with increased pressure on the government to grant more concessions for large expensive hotels inside the park.

The number of international tourists has risen annually (figure 9.3),

FIGURE 9.2 Royal Chitwan National Park has become a major tourist destination with numerous large and small hotels. (Courtesy Lori Price)

with the steepest increases from 1987 to 1988 and since 1996. Government figures show that during the 1995 tourist season 64,749 people visited Chitwan, a 24% increase from the 1990 tourist season. Of those 64,749 tourists, the largest percentages of foreign visitors to Chitwan were from India (15%), the United States (9%), and Germany (6%). The remaining tourists came from the United Kingdom, the Netherlands, Japan, France, Taiwan, and a number of other countries. In 1996, at the beginning of the most recent boom, almost 84,000 tourists visited Chitwan — and more than half (46,610) came from industrialized nations. Tourists who visit Chitwan typically account for about 20% of all tourists visiting Nepal and 75% of all tourists visiting national parks. A striking feature in 1997 and 1998 was the rapid rise in the number of Nepalese tourists, probably a reflection of a small but growing middle class. In 1988 the idea of indigenous tourism seemed rather ludicrous.

According to information collected from the 1994 employee survey (appendix A), the average wage of villagers employed by the hotel industry was approximately $28 (SD = $24) per month, or $336 per year. These figures were slightly lower yet similar to those quoted by hotel managers, who said their employees earn approximately $32 per month, or $384 per year

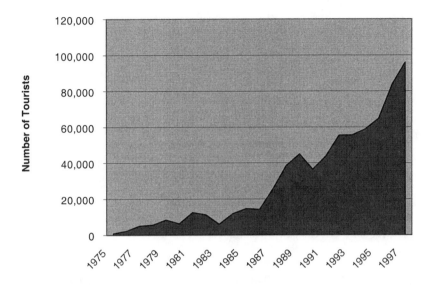

FIGURE 9.3 International visitors to Royal Chitwan National Park, 1975–1997. (Courtesy Nepal Department of National Parks and Wildlife Conservation)

(table 9.2). Employees in the Chitwan ecotourism industry earned approximately 2.5 times the average annual income in Nepal for 1994.

The Chitwan hotel industry, which included the forty-nine hotels in operation in 1994, generated an estimated total revenue of $4.5 million. The capacity of local communities to capture this revenue, however, has been limited. Nonlocals, either Nepalese from outside the Chitwan District or foreigners, owned 61% of the hotels in 1994. The hotel industry in Chitwan employed approximately 1,100 villagers, only 1% of the district's total working-age population. In 1994 approximately 72% of employees in the hotel industry were locals, originally from the Chitwan District, but fewer than 2% were women.

The package-tour rates for Chitwan attract bargain-seeking tourists; 55% of tourists in 1994 stayed at low-budget cottages and medium-price hotels with an average package tour price of $66 (SD = $51). This includes a two-day stay but not the park entry fee or tickets for boat rides, jungle drives, or other diversions. The two-day park entry fee in 2001 was $13 for foreigners, while Nepalese tourists pay only a nominal fee. However, most low-budget cottages, which constitute at least half of all lodging, often engage in such fierce competition that they offer rooms far below the going

TABLE 9.2.   THE ROYAL CHITWAN NATIONAL PARK ECOTOURISM INDUSTRY,
HOTEL EMPLOYEE WAGES AND TOUR COSTS, 1994 TOURIST SEASON

Hotel type	Number of hotels	Number of employees	Average employee salary/month (US$)	Package price* (US$)
Low budget	28	253	21.91	0–49
Medium price	10	74	29.96	50–99
Expensive	11	657	41.04	100–230
Total	49	1,084	92.91	

NOTE: Data based on hotel employee survey of 1994. The categories listed under type were based on surveys of packages offered by hotels in the area.
* Package price paid by tourists for food, lodging, and tourist activities for two days and three nights. The hotels were classified according to package rates.

rate of about $10 per night. This is possible because most lodges make their highest profit from the restaurants attached to each lodge.

Seven tourist lodges have concessions to operate inside the park boundary, and all are expensive, beginning at $100 a night. Almost all operate seasonal tent camps, which allow guests to spend a night deeper inside the park. The seven lodges maintain their own elephant stables for jungle rides, while the less expensive lodges must rent space on government elephants. Most tourists at these seven lodges book their stay through overseas travel agents. Several expensive lodges are owned jointly by wealthy Nepalese and foreign business interests; none is owned by local residents of Chitwan.

Budget-minded tourists can enjoy the same wildlife experiences as those who stay at the more posh tourist lodges if they know where to go in the park and line up a good nature guide. Of the 119 tourists from industrialized nations surveyed in our study, 39% said they would be willing to pay more to enter the park, 52% said they would not, and 9% were uncertain. Amenable tourists were willing to pay, on average, an extra $6.50 (median = $8, SD = $3.50) to enter the park.

The Earning Capacity of Nature Guides

Promoters of ecotourism typically tout the training and employment of local nature guides as one of the many spin-off benefits. However, the opportunity to earn good salaries often attracts enterprising individuals from outside the target area of the ecotourism program, thus diminishing the benefits for locals. One hundred four, or 74%, of listed nature guides for

Chitwan were permanent residents of the Chitwan District. Twenty-six percent had moved to the area within the past five years from other districts in Nepal and India. For junior nature guides, the average monthly salary before nature guide training and certification was $16. After participating in the nature guide training program and receiving certification, their average monthly salary increased by 36%. The average monthly salary for senior nature guides in 1994 was $29 a month, a 29% increase from wages earned before senior guide training and certification. As in the hotel survey, only 2% of the nature guides in the survey were women.

The Effect of Ecotourism on Household Income

To assess the effect of ecotourism on the income of villagers living near Chitwan, our research team randomly surveyed 996 households in seven of the thirty-six villages adjacent to the park. Only 44 households, 4% of the houses surveyed in the study area, reported having family members directly employed in the ecotourism industry. The average monthly salary of those employed by the ecotourism industry was $41 — similar to but slightly higher than wages reported by hotel employees. Another 2% of the households in our survey earned money from the sale of products or the provision of services associated with ecotourism. The average amount of direct and indirect income per household, for the 6% of households affected by ecotourism, was approximately $50 per month, or $600 per year. Of this $50, $34 came from direct employment in ecotourism.

The distribution of economic benefits to households from ecotourism was limited geographically. The financial effect of ecotourism on households decreased dramatically with distance from Sauraha (table 9.3). Even within the Bachhauli Village Committee, the households affected by ecotourism were concentrated primarily in the Sauraha ward. However, other wards adjacent to hotels did not receive the monetary benefits we had expected. Probably, other wards lack the capital to start up small lodges. We did not sample the buffer zone directly across from several hotels in the far western end of the park. However, we feel confident that the same low level of household penetration of ecotourism dollars would hold.

Shortcomings of the Privately Owned Ecotourism Business

Our findings show that until 1996, the distribution and amount of economic benefits to local communities was limited. The employment poten-

TABLE 9.3.   SUMMARY OF DIRECT AND INDIRECT MONTHLY HOUSEHOLD
INCOME FROM ECOTOURISM IN CHITWAN STUDY AREA

Village committee	Number of houses surveyed	Houses affected (%)	Indirect income (US$)	Direct income (US$)	Total income (US$)	Household income[a] (US$)	Distance from Sauraha[b] (km)
Bachhauli	179	22 (40/179)	494	1,910	2,404	60	0
Kumrose	180	6 (11/180)	290	80	370	34	0–1
Kathar	180	1 (2/180)	0	86	86	43	2–3
Khairahani	45	2 (1/45)	24	0	24	24	3
Patihani	135	3 (4/135)	28	4	32	8	3
Bhandara	137	2 (3/137)	100	20	120	40	3–4
Piple	140	1 (1/140)	18	0	18	18	>4
Total	996	6 (62/996)	954	2,100	3,054		

[a] Average monthly household income = total income – number of houses that receive indirect or direct benefits from ecotourism (households affected by ecotourism).

[b] Calculated as distance from park entrance to midpoint of village development committee.

tial in ecotourism is low and seasonal (some hotels reduce staff during the monsoon season or shut down entirely). The direct economic effect of ecotourism on household income is marginal. Further, the indirect effect of ecotourism on household income is virtually nonexistent; few households reported receiving money from the sale of products or provision of services related to ecotourism. This lack of profits reflects minimal market diversification from this macroenterprise.

Advance bookings made in other countries or in the capital city of Kathmandu siphon from the local economy some profits generated by the hotel industry. We estimated that 54% of hotel reservations are booked and paid for in advance, outside Sauraha. Until relatively recently, most hotels rarely purchased food grown locally, preferring to fly or drive it in from Kathmandu. Moreover, the preponderance of low-budget ecotourist hotels, highly discounted package tours, and inexpensive park entry fees in Chitwan means that ecotourism is undervalued (Wells 1993).

Owners of the largest hotels might view the employment of more than 1% of the Chitwan workforce as an important contribution. But with 240,000 people living adjacent to the buffer zone, it is hard to view ecotourism, as practiced until 1996, as a tool for conservation. It is clear that if incentives fail to reach the surrounding communities, they have no reason to view the park in a positive manner so that they refrain from collecting firewood, starting fires, or poaching rhinos and tigers.

Green Ecotourism

Much of the data presented in this chapter was first published in the journal *Conservation Biology*. The article (Bookbinder et al. 1998) launched a strongly worded response signed by representatives of all the expensive tourist hotels operating in Chitwan. The hoteliers provided no information to refute the evidence collected from the survey of nearly 1,000 households. Rather, they sought to portray the expensive hotels as contributing to conservation in Chitwan, in part because they pay an annual concession fee of $20,000 per hotel. The low-budget lodges are exempt from this fee. The truth is that the expensive hotels are required to pay the concession for operating within the park boundary or they would lose their government license to do business.

This situation thrusts the idea of green ecotourism, or certified tourism, into a new perspective. Several conservation groups, including the Rainforest Alliance and World Wildlife Fund, have promoted transforming the market for ecotourism by certifying enterprises that abide by best practices: promoting ecologically sensitive tourism, employing local people, reducing waste, and recycling a significant amount of profits to local communities. The rationale is that consumers will patronize tourist operators that are certified instead of those that are not if tourists are educated about the choices available. Certainly, creating a green ecotourism industry in Chitwan would be an important way to link the economics of privately based ecotourism with conservation. The next chapter explores other approaches.

Mounting pressures on natural resources in developing nations in Asia make conserving lands adjacent to protected areas an important goal, especially where populations of large mammals are a primary focus for conservation. In Chitwan, however, the privately owned ecotourism industry did not employ locals or buy locally made products at a scale sufficient to make a noticeable difference. Thus it was unable to influence local support for conservation. As this chapter demonstrates, the private ecotourism industry per se differs little from more exploitative forms of tourism and is unlikely to recycle significant amounts of money, which would change local attitudes toward conservation. New approaches to ecotourism are therefore essential.

CHAPTER 10

MAKING ROOM FOR MEGAFAUNA:
PROMOTING LOCAL GUARDIANSHIP
OF ENDANGERED SPECIES AND
LANDSCAPE-SCALE CONSERVATION

The answer is an eco-development project. What's the question?
—Barry Coates, conservationist, critiquing foreign aid programs
for conservation

IN THE SPRING OF 1988, I took one last walk along the Rapti
River, the boundary between Chitwan National Park and the settlements that partly surround it. In two days I would be on a plane,
heading back to the United States to finish my postdoctoral fellowship with the National Zoological Park. As I approached the village
of Bagmara, I watched as a forlorn pair of emaciated cows crossed
the Rapti and ambled toward home (figure 10.1). They had obvi-
ously spent the morning grazing inside the national park. The scene
disturbed me. The livestock's unchecked grazing of habitat set
aside for rhinoceros and other endangered species symbolized the
gradual degradation of a park that I had grown to love. I felt com-
pelled to return to Chitwan, to do something meaningful to preserve its habitat and protect its wild residents. For better or worse,
the conservation of rhinoceros and other Asian megafauna had become my mission in life. But after returning to Washington, D.C., I
had little luck with several grant proposals. Few donors supported
the traditional approach to conservation, which concentrates on

FIGURE 10.1 Emaciated free-ranging cows wander back into the buffer zone after grazing inside Royal Chitwan National Park, 1988. The photograph epitomized my despair for the future of Chitwan and its rhinoceros. (Photo by Eric Dinerstein)

the expansion of protected areas and effective patrolling by armed guards. Instead, a new kind of conservation intervention was gaining greater enthusiasm—the eco-development project, or integrated conservation and development project.

By the early 1990s, conservation had moved sharply away from strict protection in favor of meeting the basic needs of local people. Most money has flowed into expensive eco-development projects that attempt to integrate biodiversity conservation with rural development. By 1996, this trend was so pronounced that some conservation nongovernmental organizations (NGOs) earmarked more than half their money for people-oriented conservation programs instead of strict protection efforts. The World Bank alone pours millions of dollars into such projects all over Asia (MacKinnon, Mishra, and Mott 1999; Wells et al. 1999). Because of this trend, and despite precipitous declines of rhinoceros, tigers, elephants, and other Asian megafauna throughout much of their range, the protection or expansion of parks and reserves has received only marginal funding. Conservation NGOs also have neglected antipoaching information networks and other measures that have proved to be effective in recovering decimated populations.

Why have eco-development projects become so popular among foreign aid agencies? First, these projects have rural development components, which these donors are familiar with. Second, aid agencies view traditional approaches to conservation — often described as the "guns and fences," "preservationist paradigm," or exclusionary approach — as ineffective (Terborgh 1999). The pervasive feeling among some donors is that excluding people from an ecologically sensitive area only delays the inevitable — they eventually will overrun or encroach heavily on the reserve. Yet a recent study shows that strict protection of areas is effective at reducing habitat degradation (Bruner et al. 2001). Other new studies are beginning to show that the near abandonment of traditional approaches, together with the heavy emphasis on eco-development projects, also has failed to conserve endangered large mammals or other elements of biodiversity. A review of eco-development projects in Indonesia concluded that these projects failed to meet either of their twin goals: biodiversity conservation and local development (Wells et al. 1999).

Despite their complexity and other problems, eco-development projects still have a critical role in defining the future of biodiversity in developing nations (Oates 1999). In Asia most reserves are too small to maintain viable populations of large mammals over the long term. Yet the landscapes surrounding them are densely settled. Eco-development projects may be an important tool for conserving landscape features such as corridors, buffer zones, and multiple-use areas that enhance the persistence of endangered species living in fragmented habitats or small reserves. But these projects require a careful design and certain preconditions.

This approach has succeeded in Chitwan. The Chitwan buffer zone became the target of an eco-development project that was designed to mesh with an existing strict preservationist program that had been developed for the core area of the park. This has produced favorable results for rhinoceros, tigers, and nearby human communities. The new national legislation mandating community forest management and recycling of park revenues to local communities guaranteed the long-term sustainability of these revenues. Lobbying for this legislation was an essential component of the general conservation program.

The Evolution of the Eco-development Component

Incentives: Perceived or Real?

To date, Western conservationists attribute much of the recovery of Chit-wan's wildlife to perceived economic incentives. However, as the discussion in chapter 9 illustrates, private ecotourism ventures do not necessarily re-duce threats to populations of endangered species. Another misconception involves permission to cut thatch grass inside the park. In Chitwan, most villagers use thatch grass as roofing material (figure 10.2). A long-standing policy allows villagers access to the park's grasslands to cut thatch grass (*Imperata cylindrica*) and canes (*Narenga porphyracoma, Saccharum* spp.) dur-ing a two-week period every winter. Western conservationists have touted the thatch grass–collection program as an example of positive people–park interactions. Proponents of the program contend that without the park, vil-lagers would have no place to collect thatch and canes. The truth is much more complicated. First, plant succession in Chitwan's grasslands has greatly reduced the amount of thatch species available (Lehmkuhl 1989). Second, instead of collecting grass, many people conceal firewood, their real target, in their grass bundles during the thatch access period (John Lehmkuhl, personal communication, 1986). Since 1990 villagers have been selling much of the grass cut inside Chitwan to contractors who, in turn, sell it to a local paper mill. The local villagers receive a meager return for their labor (Arun Rijal, unpublished data, 1999). But the sale of thatch grass does not provide strong enough incentives to elicit the support of villagers for landscape-scale conservation.

We faced some other hard facts at the beginning of our project in 1994. First, the true reasons for the recovery of the rhinoceros and tiger popula-tions were strict protection of the core breeding population of these species and their habitats inside the park as well as the law-abiding nature of Nepalese citizens living around the park boundary. Relying on strict pro-tection alone, however, was a temporary solution. Second, the park was too small to maintain viable populations of tigers, rhinoceros, or elephants. Third, domestic livestock was severely degrading wildlife habitats in the buffer zones and corridors attached to the park. Fourth, locals had little in-centive to change their patterns of using natural resources that negatively affected conservation targets. Much of the buffer zone around Chitwan was

FIGURE 10.2 For two weeks in late winter, local people may enter Chitwan to cut thatch grass and canes to use as building and roofing materials. (Courtesy Tom Kelly)

in worse shape in 1994 because of human encroachment than it had been when the rhinoceros field project started in 1984. Finally, the government had made little effort to make local people partners in planning efforts. How could we expect habitats and buffer zones to be restored, dispersal corridors to be maintained, and core areas to remain inviolate without the good faith and collaboration of local villagers?

The Project Design

Although the ultimate goal was to provide a series of incentives to promote local guardianship of endangered species and their habitats, the initial strategy of the eco-development project was to take the pressure off the natural ecosystem by meeting the resource needs of local people through better management of buffer-zone forests. Demand for firewood, fodder, and grazing areas has always put severe pressure on the habitats of Chitwan and its buffer zone. Initial investments in timber, fuelwood, and fodder plantations addressed some of these basic needs (figure 10.3). Beginning in 1994, a second layer of investments promoted community-based ecotourism, profit sharing, alternative energy sources, and improvement of livestock. Subsequently, we have seen less pressure on park resources, regeneration of riverine forests, and regrowth of critical grasslands and aquatic habitats (figure 10.4). A main mechanism for these improvements was recycling park revenues, income that became available in early 1998.

Project Ingredients, Activities, Benefits, and Effects

A $10,000 award from the U.S. Agency for International Development (USAID) in 1988 financed the creation of a native tree nursery — the precursor to a community forestry program — on the private land of Shankar Choudhury, a forest ranger with the King Mahendra Trust for Nature Conservation (figure 10.5). The following year, foresters fenced and established a locally managed tree plantation on a degraded 32-ha plot of government land in the buffer zone (hereafter, Bagmara Community Forest). They planted native rosewood (*Dalbergia sissoo*) and khair (*Acacia catechu*) as well as three other native tree species (*parke, Albizia lucida; neem, Melia adzirachta;* and teak, *Tectona grandis*) to provide timber and firewood. They planted some tree species, particularly rosewood, along riverbanks to pro-

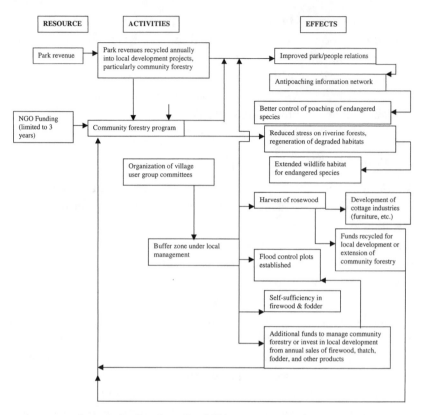

FIGURE 10.3 Conservation benefits to Royal Chitwan National Park as conceived in 1994. The original design saw "extended wildlife habitat for endangered species" as an endpoint, with no anticipation that the project would lead to community-based ecotourism in the buffer zone. Funding came from the King Mahendra Trust, the World Wildlife Fund, and the Biodiversity Conservation Network.

mote flood control or provide flood protection. These other species are an incentive for locals to support habitat regeneration over the long term because the wood of these trees is valuable, likely to bring large sums as trees are harvested over twenty years (table 10.1). As native species, they serve a second function as valuable wildlife habitat.

This community forestry program produced a dramatic shift in local attitudes. Initially, many people were antagonistic. Some even tried to tear down the fences. They saw the program as restricting the movements of their free-ranging cattle. After one growing season, however, the recovery of valuable thatch grass to serve villagers' needs, and the opportunity to harvest it, convinced skeptics that plantations were useful. Three years later, the

A

B

FIGURE 10.4 Initial steps in habitat recovery. (A) Free-ranging cattle graze in front of a fenced buffer-zone regeneration area. Regrowth in the fenced area (excluding emergent trees) is after only one year of protection. (B) Another, nearby location shows a few cattle grazing in front of a nine-year-old *sissoo* plantation. (Photos by Eric Dinerstein)

FIGURE 10.5 Shankar Choudhury, a forest ranger with the King Mahendra Trust, was able to convince villagers to set aside and manage the buffer zones as fuelwood, timber, and fodder plantations. (Photo by Eric Dinerstein)

saplings had shaded out the thatch grass, but they offered a more coveted replacement — a three-month supply of firewood for residents of the surrounding area as they trimmed the lower branches.

Key legislation adopted in 1993 sanctions the organization of user group committees to manage buffer-zone forests. Locally elected, the members of the committees represent various community interests. Once established, a committee petitions the local district forest officer for the right to manage buffer-zone land. Previously, only the Department of Forestry — a centralized and financially limited agency unable to stem the degradation of important forested areas — managed buffer-zone forests. Since 1998, committees have petitioned the Department of National Parks and Wildlife Conservation, which assumed responsibility of the buffer zone from the Department of Forestry.

The amount of money returned to the local community from the rosewood, combined with thatch grass and pole extraction, is expected to have yielded about $62,000 per hectare, or $1.98 million from the original 32 ha (based on the 1998 price for rosewood), by the end of the twenty-year harvest (table 10.1). The first thinning of these plantations occurred after six

TABLE 10.1. BENEFITS ACCRUING FROM TIMBER PLANTATIONS ESTABLISHED IN 1988

Benefits	Amount harvested each year	Harvest cycle	Income (US$)
Thatch grass	2,813 kg dry weight	First 2 years only	225
Thinning of poles	1,250 kg/ha	After 5 years	4,960
Thinning of poles	625 kg/ha	After 10 years	14,880
Thinning of poles	250 kg/ha	After 15 years	11,904
Harvest of all trees	375–400 kg/ha	After 20 years	29,762
Total revenues/ha over 20 years			61,732

years when the villagers got around to it (figure 10.6). In contrast, the same hectare of land placed under cultivation (rice, wheat, and mustard) would return only about $6,000, less the cost of fertilizers and seeds, over the same twenty-year period.

To have even a marginal effect on park protection and natural resource management in the buffer zone, the program had to accelerate quickly. Between 1994 and 1996, through grants from the now-defunct Biodiversity Conservation Network, an arm of the Biodiversity Support Program of

FIGURE 10.6 Thinning the rosewood plantations after five years realized profits of nearly $5,000 for local villagers. (Courtesy King Mahendra Trust for Nature Conservation)

USAID, the local user group committee expanded the Bagmara Community Forest regeneration area to 460 ha. During the same period, villagers planted native trees and allowed the natural regeneration of another 1,050 ha of degraded riverine forest and alluvial grassland in the Kumrose buffer area (only about 20% of this area in Kumrose was plantation; the remainder was natural regeneration). In 1997 a grant from the Save the Tiger Fund helped fence 250 ha for natural regeneration in the Kanteshwari buffer area, to the east of the park. Plans for additional extensions are under way in several other areas of the buffer zone. By 1999, the community forestry program had brought more than 30 km^2 under local management. Support from other NGOs helped establish another 20 km^2 of *sissoo* plantations in the eastern Chitwan buffer-zone areas to promote flood control. In our study area, the program spent relatively little money to establish native tree plantations of *sissoo* because of a deliberate decision to encourage communities to concentrate instead on natural regeneration across much larger areas. Natural regeneration is far less costly, and a wider range of wildlife species prefers this habitat to native tree plantations, even though monotypic stands of *sissoo* plantations mimic what occurs naturally. *Sissoo* is one of the first trees to colonize floodplain islands in Terai rivers.

Fencing off regeneration areas brought an unexpected bonus — the fence formed an effective barrier between the park's large herbivores and croplands. Six strands of barbed wire strung along cement poles bordered by deep trenches greatly reduced nightly forays by large crop-raiding herbivores. Villagers who farmed land adjacent to regeneration areas reported much less crop damage and higher yields than did villagers who farmed up to the edge of the park (Rijal, unpublished data, 1996). Access gates to regeneration areas adjacent to the park boundary (the Rapti River) remained open at night to allow free movement of rhinoceros and other wildlife. Rhinoceros using these areas adapted quickly to the gate system.

Flood control turned out to be another significant benefit of regeneration. In 1993 a catastrophic flood caused tremendous damage to the Chitwan area, creating many refugees and burying large expanses of farmland under 1 m of sand. But even villages in the hardest-hit areas were spared where remaining strips of riparian buffer-zone forest deflected the force of the floods. Although a tragic episode, the 1993 flood convinced local leaders that retaining riparian forest was essential for the safety and economic well-being of agriculturists living close to the park. The flood also accelerated efforts to relocate an enclave of villages in the Padampur area to another location much farther from the river. However, the most dramatic conservation achievement was the recolonization of regeneration areas by endangered wildlife. These new inhabitants became the focus of a community-based ecotourism program.

Community-Based Ecotourism and Biodiversity Conservation

By 1989, many conservationists and local villagers had recognized that tourism in Chitwan was predominantly a private enterprise controlled by a handful of operators. Profits for surrounding villages were marginal. To return more of the profits to local groups, the Bagmara committee (representing more than 580 households in the Chitwan buffer zone), the World Wildlife Fund, and the Biodiversity Conservation Network designed and implemented a plan in 1994 for community-based ecotourism in the regenerating forest (figure 10.7). The program established nature trails for elephant-back safaris, carefully regulated access and trail use, and hired guards from the local community to protect the area from poachers and

FIGURE 10.7 Bishnu Prasad Aryal, a local village leader, stands before a map of the buffer zone managed by the Bagmara Village Committee. The multiple uses of this zone include fodder and timber plantation, natural regeneration, thatch grass regeneration, strict protection of wildlife, and wildlife viewing for ecotourism. (Courtesy Tom Kelly)

FIGURE 10.8 A grant from the Biodiversity Conservation Network facilitated the construction of two wildlife-viewing towers that also accommodate tourists overnight. (Courtesy Tom Kelly)

trespassers. The villagers built a wildlife-viewing tower (*machan*), which led to overnight jungle-camping for tourists, who stay in the tower (figure 10.8) and observe rhinoceros using wallows. The local committee took over the management of the tower, and local people living next to the buffer zone have experienced a tremendous sense of empowerment from managing natural areas and resources. The plantations meet most of the local demand for fodder grasses and much of that for firewood. Successful community-based ecotourism reflects a growing sense of entrepreneurship.

Measuring the Conservation Effect and Economic Impact of Eco-development Projects

An Approach to Monitoring

Eco-development projects often lack the framework, funds, and staff necessary to determine whether they conserve endangered species and habitats. The framework presented here measures project effectiveness at three levels: the status of sensitive indicator species; the integrity of habitat and the landscape; and the maintenance of critical ecological processes. We applied this framework in our study of Chitwan and certain regeneration areas.

SENSITIVE SPECIES. The variables were densities of rhinoceros, densities of tigers, relative abundance of tiger prey, recruitment of both species in regeneration areas of the buffer zone, and poaching levels of rhinoceros and tigers. We also assessed the use of regeneration areas by rare forest birds and all forest birds, although these data are not presented here.

HABITAT AND LANDSCAPE INTEGRITY. The variables were regeneration of critical habitats (core areas) preferred by endangered species (tigers, rhinoceros, hog deer, hispid hare), local guardianship and improved management of habitats restored by committee activities in the buffer zone, development and implementation of plans to expand the conservation effect beyond protected areas to a landscape scale, and management of prime riverine habitats. (For alluvial grasslands and riverine forests, an area of 400 km^2 of prime habitat is considered to be a minimum-size conservation landscape to support a viable population of rhinoceros for at least 100 years.)

ECOLOGICAL PROCESSES. Most studies never measure the effects of project activities or interventions on maintaining or restoring ecological processes. Yet these processes may be the most important variables because they determine the structure and function of natural communities. These include dispersal of juveniles and adults, predator–prey interactions, and plant succession. The variables were the intactness of wildlife corridors as a proxy for measuring dispersal of tigers, predation by tigers in regeneration areas, and the maintenance of early successional habitats containing the full suite of native species.

The Conservation Effect on Sensitive Species

Rhinoceros densities have increased steadily in the regeneration area since the inception of the buffer-zone project in 1994. Between 1984 and 1988, rhinos were relatively common (Dinerstein and Price 1991), but habitat degradation, caused by intensive cattle grazing and firewood collection, reduced rhino numbers to only a few individuals in buffer areas between 1988 and 1994 (Bishnu Lama, personal communication, 1995). Beginning in

1995, we used photo-identification techniques to catalogue individual rhinos that were using the regeneration areas. From March 1998 through November 1999, an average of 2.9 rhinos used the buffer zone in Bagmara each month, whereas 8 were using Kumrose, and 19.4 were using Icharni (the control plot inside Chitwan) (figure 10.9). These numbers correspond to an average density in Bagmara of 0.7 individuals per km^2, 1.3 per km^2 in Kumrose, and 6.5 per km^2 in Icharni.

Between October 1995 and March 1997, the number of photo-registered rhinoceros using Kumrose was twenty-three; nineteen used Bagmara, and fifty used Icharni (figure 10.10). Recruitment of rhinoceros in the regeneration areas was excellent. Twenty calves were born and survived in the regeneration areas during the five-year monitoring period (1995–1999); twenty-seven were born in Icharni (figure 10.11).

A tiger census conducted in Chitwan and the adjacent Parsa Wildlife Reserve in 1996 estimated a population of 118 (Nepal 1996). Tigers in Chitwan Valley tend to live inside the park, but adjacent to buffer-zone areas or those soon to be under regeneration. Tigers, like rhinos, responded well to the increased protection of the buffer zone. None of the tigers used the regeneration areas in Bagmara in 1994 and probably did not in the years before management of the area began (Lama, personal communication, 1996). In 1995 an adult male tiger occasionally visited Bagmara (figure 10.12). From 1996 through March 1997, five tigers regularly used Bagmara and Kumrose: one adult male, two subadults, and a female and cub.

As I discussed in chapter 5, a critical component of recovery programs for endangered large mammals is reducing human-induced adult mortality, most often caused by poaching. Recovery of rhinos and tigers in the buffer zone is impossible if poaching exceeds recruitment. Poaching data for rhinoceros and tigers are the best indicator of the degree of illegal hunting in Chitwan. Between 1990 and 1997, poachers killed forty-five rhinoceros in and around Chitwan (World Wildlife Fund 1997) (figure 10.13a). Poaching of rhinoceros peaked in 1992 (24 incidents) during a period of social disruptions that weakened the antipoaching information network. After the government restored order and reinvigorated the network in 1994, poaching became insignificant—typically one or two individuals poached per year. Between 1990 and 1997, two areas in the eastern part of Chitwan saw eleven known incidents of tiger poaching (figure 10.13b). One tiger was poached in 1998.

Any monitoring program should include spatial as well as numerical

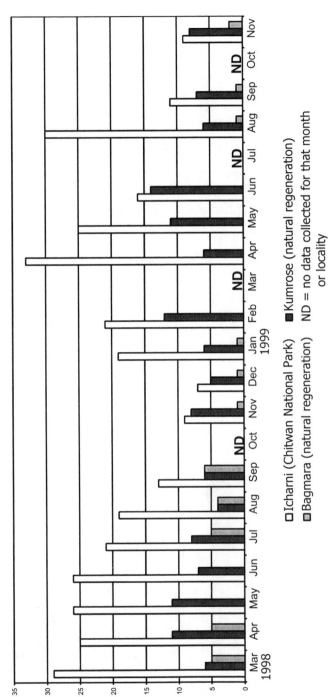

FIGURE 10.9 Total number of rhinoceros observed monthly in regeneration areas and Royal Chitwan National Park, March 1998–November 1999.

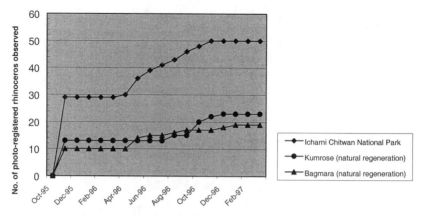

FIGURE 10.10 Photo-registered rhinoceros observed monthly in regeneration areas and Royal Chitwan National Park, October 1995–March 1997.

observations of poaching pressure. Analysis of the spatial association of army guard posts and the park boundary and seven years of poaching data for rhinoceros (figure 10.13a) reveals that

- Eighty percent of all poaching incidents were concentrated in two very small areas, even though rhinoceros range widely along the Rapti and Narayani Rivers.

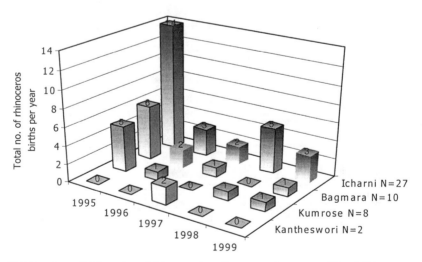

FIGURE 10.11 Total number of rhinoceros born in regeneration areas and Royal Chitwan National Park, 1995–1999.

FIGURE 10.12 An adult male tiger photographed by infrared camera trap in the Kumrose regeneration area. (Courtesy King Mahendra Trust for Nature Conservation and Nepal Conservation, Research, and Training Centre)

- Fifty percent of all poaching incidents occurred in two areas at the northern edge of the park within a total area smaller than 54 km^2.
- Most poaching incidents occurred north of major rivers where the Nepalese Army is not responsible for controlling poaching (the area is outside the park) and where no guards are posted.
- No poaching has occurred since 1995 in or around the regeneration areas.

Surprisingly, the locations of rhinoceros- and tiger-poaching incidents during the seven-year period show little overlap. I believe that this is so because tigers are territorial and rhinoceros are not.

The Conservation Effect on Habitat Integrity and Landscape-Scale Restoration

By June 1999, more than 30 km^2 of critical riverine forest habitat in the Chitwan buffer zone had recovered. The early success of local management and guardianship of these habitats is striking in many ways. For example, staffers

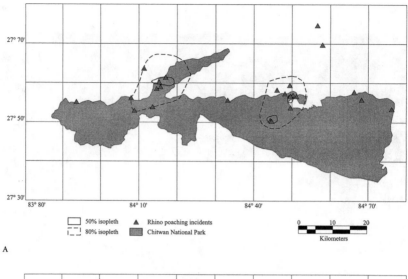

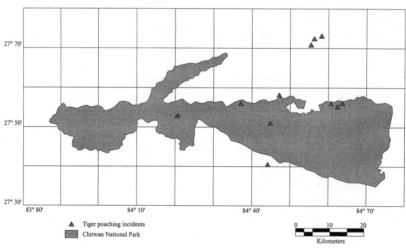

FIGURE 10.13 Poaching of endangered species in Royal Chitwan National Park, 1990–1997. (A) The forty-five incidents of rhinoceros poaching were somewhat clustered. Three small areas totaling about 54 km² were the sites of half of all poaching incidents (*solid line*). Eighty percent of the poaching incidents occurred in the areas designated by the dashed line. (B) Since 1990, the poaching of tigers has been more prevalent in the eastern half of Chitwan.

from the King Mahendra Trust and I have encouraged local managers to leave standing dead trees (snags) in the forest because they provide essential roosting and breeding sites for cavity-nesting birds and mammals. Since 1995, when we made this recommendation, not a single snag has been cut in the regeneration areas, whereas snags outside these areas continue to be cut. Also, in both Kumrose and Bagmara the local committees and volunteers have dug out extensive oxbows and provided year-round water to them. Rhinos use these oxbows as wallows; thus the oxbows are convenient sites to observe rhinoceros (figure 10.14). The Bagmara oxbow was stocked with endangered mugger crocodiles, and the wetlands there are now home to several species of wading birds and waterfowl uncommon in the area before habitat management began. The villagers have fenced the tall grassland used by rhinoceros and tigers in Kumrose to keep out cattle, and they plan to expand the area fenced. Illegal firewood collection in the regeneration areas ceased through the rigorous patrols of paid local villagers. At the same time, among villages with committees, the number of residents who cut grass, collect firewood, or graze cattle inside the park has declined to almost zero.

Finally, the committees carefully manage wildlife ecotourism in both

FIGURE 10.14 Within the buffer zone, local villagers dug out old oxbow lakes and rechanneled canals to provide year-round water and wallowing sites for rhinoceros. (Photo by Eric Dinerstein)

areas. They limit elephant-back viewing in Bagmara to one-hour rides, with a maximum of five elephants using the area at a time. Elephant drivers are required to stay on trails carefully designed for tourist wildlife viewing. The committees maintain the trails, the wildlife-viewing towers, and the short-grass clearings around them and use revenues from the entry fee charged to tourists to pay for maintenance.

Diverting many ecotourism activities to the buffer zone benefits land-scape management as well. It reduces disturbance to the core area of the park, which contains most of the breeding population of tigers. The buffer-zone forest has a greater potential for ecotourism than Chitwan, because it is accessible even during the monsoon season when flooding of the Rapti River restricts entry to the park. A last advantage is that tourists who enter the wildlife-viewing area in the buffer zone must still pay the park entry fee. The park and the buffer-zone areas benefit from the revenue from this charge.

Have community-based activities enhanced rhinoceros and tiger conservation in the vicinity of Chitwan? The answer appears to be yes. The committees and the local community members that they represent view themselves as local guardians of wildlife. They are now actively protecting the areas from would-be poachers, cattle grazers, and illegal woodcutters, and no rhinoceros or tigers have been poached in areas under their jurisdiction since the inception of the project. Critical wildlife habitat amounting to 30 km^2 is well on the way to recovery. The regenerative capacity of this landscape is extraordinary because of the tremendous productivity of the Terai riverine forests and grasslands. Rhinoceros can reach densities of more than ten individuals per km^2 in alluvial grasslands, and tigers in these habitats achieve some of the highest densities recorded in Asia. When protected from domestic livestock grazing, rhinos and tigers can reach these densities in one or two years. Furthermore, the amount of habitat regenerated in the buffer zone is now approaching, and even exceeding, the size of the smallest rhinoceros sanctuaries in northeastern India: Pabitora, 17 km^2; Khatarniaghat, 20 km^2; Laokhowa, 70 km^2; Orang, 76 km^2; and Gorumara, 79 km^2 (Foose and van Strien 1997).

These regeneration successes — as well as the generation of profits — have not gone unnoticed. Several other committees along the periphery of Chitwan are determined to use the model provided by Bagmara and Kumrose. One recently planned regeneration area will form key habitat in the last remaining forested corridor, connecting the deciduous hill forests of the Siwalik Range to the Mahabarat Range at the foothills of the Himalayas by way of Royal Chitwan National Park. The Barandabar forest, a 40-km^2 block in the southern part of this corridor, still has resident rhinoceros and tigers. It also contains one of the most important wetlands in South Asia (called Bis Hajaari Tal, which means 20,000 lakes), an area used by many migratory birds and a once-dense population of rhinoceros (figure 10.15). The regeneration project will maintain dispersal corridors for tigers and many other species between the Bis Hajaari Tal area and Chitwan, a distance of 10 km. Approximately 80% of the corridor would become part of the national park. The other 20% would consist of two buffer strips managed by local committees. The Save the Tiger Fund has been a steady supporter of this initiative, and the Global Environmental Facility and the United Nations Foundation recently awarded $1.5 million to the King Mahendra Trust for Nature Conservation.

Because cattle grazing in lowland habitats must be reduced before any significant improvement in the conservation landscape can occur, the third phase of the field project will experiment with improved breeds of livestock. If local villagers have access to veterinary care for livestock, they will accept improved breeds. Improved breeds are stall-fed with fodder harvested from adjacent plantations (these animals are too valuable to be allowed to roam freely). The total population of free-ranging domestic stock will decline, because farmers will sell off unproductive local cattle once they see the bene-

FIGURE 10.15 Bis Hajaari Tal is a regionally important wetland that offers great bird-watching and excellent views of the Himalayas. (Courtesy Tom Kelly)

fits of stall-feeding. Stall-feeding cattle also reduces farmers' conflicts with wildlife. The threat of predation by tigers living or dispersing along the borders of the buffer zone will disappear.

The promotion of stall-feeding livestock has another ecological benefit. Villagers deposit cattle dung in methane gas digesters and process it as cooking fuel, which lessens villagers' dependence on fuelwoods. The Kumrose Village Committee alone has recently installed more than 100 methane gas digesters and plans to install more. The increasing use of this cooking fuel (biogas) provides another incentive for stall-feeding livestock. The animal dung that fuels the digesters is easier to collect from stall-fed animals than from free-ranging cattle. The slurry left in the digesters can be used as fertilizer.

The greatest hope for landscape-scale conservation is that the government will embrace an ambitious plan to support the King Mahendra Trust and other NGOs for regenerating a much larger section of the buffer zone. Of the 750 km^2 of forests and grasslands that have been officially demarcated as the Chitwan buffer zone, 58% (435 km^2) could be successfully regenerated, while the other 42% would remain under cultivation (Arun Rijal and Anup Joshi, personal communication, 2000). Assuming a conservative density estimate similar to that in the Kanteshwari forest regeneration area (where 12 rhinoceros occupy 4.5 km^2, a density of 2.7 individuals per km^2), the population of rhinoceros in the buffer zone could theoretically increase by 1,174 individuals. Factoring the recovery of other species into the equation would increase the value of this effort many fold. The number of breeding female tigers is likely to exceed fifty, which is considered the minimum threshold for short-term viability. Other endangered species, such as giant hornbills, marsh mugger crocodiles, and python, are also likely to benefit. Moreover, the regeneration program would facilitate the dispersal of large mammals, increase protection of key riparian habitats, and provide some important ecological services such as flood control.

The Conservation Effect on Restoration of Ecological Processes

One of the most challenging aspects of conservation biology is the maintenance or restoration of ecological processes. Some ecological processes, such as photosynthesis, require no intervention from humans to continue. Others, such as large predator–prey interactions, can be permanently altered by humans and difficult to restore. The loss of these interactions or processes can destabilize an entire ecosystem and may reduce the viability of many species. Monitoring these processes is critical but is rarely done. In the Chitwan context, I identified three important ecological processes to follow: restoration of successional patterns, predation by large carnivores on natural prey, and dispersal of large carnivores.

Restoration of some processes may not happen immediately, but initial trajectories are encouraging. As I described earlier, project interventions have fostered plant succession in the buffer zones, from degraded scrub to riverine forest, leading to the restoration of native vertebrate communities. The second process, predation by large carnivores, assumes that large prey species have recovered in sufficient numbers in the regeneration areas to attract tigers and leopards. To date tiger predation on native ungulates in the buffer zone is uncommon, although it has become more frequent since 1997 when tigers began killing sambar deer living in the Kumrose area. Dispersal is perhaps the most logistically challenging feature to monitor accurately. We have not been able to determine whether project activities have enhanced the dispersal of tigers, but restoration along the buffer zone most likely has had a positive effect. Restoring the Barandabar forest corridor — a known tiger dispersal corridor — and corridors identified as part of the Terai Arc program will be a better test.

Evaluation of the Community-Based Ecotourism Project

During the first year of operation, November 1995 through October 1996, 10,632 tourists visited the Bagmara Community Forest; the area realized $276,432, almost exclusively from entry fees. Before November 1995, local people and Royal Chitwan National Park earned no income when tourists used this area; local nature guides took tourists to the Bagmara Community Forest, but neither they nor Chitwan staff had the legal right or mechanism to collect a user fee.

As of late 2000, the Bagmara committee had earned approximately $100,000 from the community-based ecotourism program (Arup Rajuria, personal communication, 2001). Trust staff members and I estimate that the Bagmara committee earned about $100 per hectare annually from community-based ecotourism from mid-1998 through late 2000. This is far less than the value of harvesting rosewood per hectare amortized over twenty years, which is about $3,100 per hectare annually. However, ecotourism revenues are a steady revenue stream to the local economy. They are an excellent complementary enterprise to the rosewood plantations, which can provide cash benefits only at five-year intervals during thinning. The estimated return of $100 per hectare from wildlife viewing will increase dramatically if the local committee keeps a higher percentage of ecotourism dollars than the hotels do.

In 1997 the Bagmara committee decided to divert some profits from its successful elephant-back ecotourism enterprise to pay local fishermen not to fish in the Khageri River. This river, a tributary of the Rapti River, flows through the Bagmara regeneration area. The payoff has been dramatic, both ecologically and financially. In one year, the marsh mugger crocodiles returned to the Khageri River from the Rapti to feed on an abundant supply of fish. The committee has set up a highly profitable dugout-canoe ride guided by those who formerly fished this river. From the canoes, tourists can view and photograph crocodiles and wading birds along the riverbanks (figure 10.16). Now several other local groups want to develop similar boat-based wildlife excursions in their area along the Rapti River. Here is an example of visionary conservation activities that evolved in the local community. One concern about introducing microenterprise activities has been that if eco-

FIGURE 10.16 Endangered marsh mugger crocodiles have returned to the Khagadi River and are a major tourist attraction there. (Photo by Eric Dinerstein)

nomic benefits are not immediate, locals will quickly lose interest. But we have found that people will invest their own resources, even when there is a brief lag before the wildlife return and economic benefits are realized.

Bagmara's approach has been replicated in Kumrose, where villagers led by Dhan Bahadur Dahal (figure 10.17) now protect a regeneration area twice as large as the Bagmara site. After a dugout-canoe ride to view crocodiles along the Khageri River, and to observe rhinoceros in the regeneration areas, George Schaller of the Wildlife Conservation Society acknowledged the project by the King Mahendra Trust as "the most encouraging wildlife restoration program in Asia."

As it turns out, local people are also unconcerned about rhinoceros' occupying the rosewood plantations. We expected the rosewood plantations to be potential rhinoceros habitat but thought the people would value the trees so much that they would restrict the rhinos' access to these areas. However, rhinoceros cannot push down and browse trees that are older than nine or ten. Instead, they feed heavily in the winter months on understory shrubs, such as *Callicarpa macrophylla*.

The initial revenues generated by the various income streams of the

FIGURE 10.17 Dhan Bahadur Dahal, a progressive village leader, addresses the Kumrose Village Committee to promote the development of community-based ecotourism and nature conservation in the buffer zone adjacent to Kumrose. (Courtesy Tom Kelly)

project have financed the refurbishing of a health post, supported improved medical care, built new schools, improved village roads leading to markets, compensated farmers who have lost livestock to tiger depredation, and invested in new ecotourism ventures (such as the dugout-canoe rides to view crocodiles). The revenues lent some degree of self-sufficiency to many of the components of the ecotourism program. One of the more recent lessons is not to emphasize economic incentives that promise to raise the standard of living for individual households. A better approach is to stress improving the quality of life for communities, as revenue generated from ecotourism and forestry may go further and be easier to manage when spent on community goods and services.

Measuring Changes in Awareness

In February 1998, I revisited the Kanteshwari forest of Chitwan's buffer zone twelve years after the 1986 rhinoceros census. During the 1986 census, I had photographed all nineteen individuals using the area. By 1994, the rhinos were gone (Yonzon 1994), and cattle grazing and woodcutting had reduced the dense jungle to a highly degraded scrub. Fenced from cattle and woodcutters in 1996 by the Kanteshwari committee, the same site now holds nineteen rhinoceros. My guide through this area was a local schoolteacher who had formed a nature club for almost sixty students. Only a few years earlier, the parents of these children had viewed rhinoceros as a menace, and the children had shot birds with slingshots. Today these children and their teacher speak with pride about their wildlife neighbors. No matter how popular Asian wildlife is to visitors of national zoos in capital cities, the real defenders of Asian wildlife are the people who coexist with them.

A test for local attitudes toward wildlife came in 1998 when a tiger killed a local villager at the edge of the Bagmara regeneration area. Man-eaters are infrequent in the eastern side of the park; they are more common in the west. Naturally, these man-eaters are the worst public-relations agents for conservation. Villagers react angrily to the loss of human life and have on occasion set out poison baits. In Bagmara, staffers from the King Mahendra Trust quickly captured the man-eater (an old male) and shipped him to the Kathmandu zoo. Buffer-zone funds were used to compensate the family of the person who died. A tragedy that normally would have strained relations between the park and the local people was rapidly and sympathetically dispatched.

Prerequisites for Success

Just as the privately owned ecotourism industry in Chitwan was not a panacea for conservation, neither would I recommend applying the community-based ecotourism approach by itself. A combination of the two would be the ideal. A distinct set of conditions is essential for successful implementation of a community-based approach. These include a relatively accessible wildlife reserve with a well-protected core area that contains a charismatic and visible species or assemblage of species; an uncultivated buffer zone where regeneration and environmentally friendly economic activities are options; a secure land-tenure system to minimize in-migration in response to the magnet effect of eco-development projects; a stable, privately owned ecotourism industry that can serve as a precursor to a community-based approach and absorb some initial costs (or a benevolent entrepreneur willing to help local leaders); national policies that enable locals to participate in enterprise activities in buffer zones adjacent to protected areas; a cooperative working relationship between local people and officials of the protected area; and strong local institutions that enforce traditional conservation rules, ensure the equitable distribution of benefits to communities from joint activities, and respond to changing economic conditions and new opportunities with a community mind-set.

Enabling Policies: New Legislation for Strengthening Linkages Between Ecotourism and Megafauna Conservation

An even more powerful tool than community-based activities is available for megafauna conservation, if the political will to reinforce it exists. Nepalese conservationists began lobbying to allow local village committees to take over the management of degraded Department of Forestry lands adjacent to protected reserves. Nothing was done until 1993, when a major reform in national policy allowed the creation of legal buffer zones around existing protected areas. Additional landmark legislation came in 1995, when parliament ratified a series of by-laws. One law required that 50% of revenue generated in protected areas be recycled into local development programs, instead of being returned to the Ministry of Finance. The recycled revenues program began in March 1998, paving the way for legal economic incentives to reduce pressures on core reserves and to conserve wildlife habitats outside parks. These landmark policy decisions are changing the face of conservation in Nepal. By 2001, they were generating as much as $400,000 a year for community activities within the Chitwan buffer zone. This revenue equals, and in some cases surpasses, what many government agencies or foreign donors might spend annually on rural development projects. This, in turn, frees those sums for donation to committees that, unlike Bagmara, lack a buffer zone area for community-based ecotourism.

Unsolved Problems

Despite innovative legislation, ecotourism in Chitwan remains a tourists' market; most visitors pay less than $20 for their wildlife experience, and low-budget hotels capture only a small portion of potential profits. The government has not restricted hotel construction outside the park or limited the number of visitors to the park, and engages in only minimal tourism planning and management both within and along the periphery of the park. Few hotels run at full capacity, there are no minimum prices set for tourism activities (except park entry fees), and the majority of hotels offer highly discounted package tours. Correcting these flaws would only enhance the significant efforts already under way to forge a stronger link between conservation of megafauna and economic benefits for local stakeholders. Establishing a green tourism certification program in Nepal, as I mentioned in chapter 9, could go a long way to address these problems.

Where Are Eco-development Projects Most Likely to Succeed in Conserving Habitat for Megafauna?

A Model for Siting Eco-development Projects

It is evident from the many lessons learned in implementing eco-development projects throughout Asia that they are not suitable everywhere. Based on the experiences from Chitwan and from my own observations at other sites, I have developed the following model for predicting where eco-development projects targeting megafauna are appropriate and likely to succeed (figure 10.18). The three axes of the graph represent three determining variables of megafauna conservation in Asia: (low-high) habitat integrity, including the relative resilience of the main habitat type; (high-low) poaching pressure on rhinoceros, elephants, and tigers as well as tiger prey and the

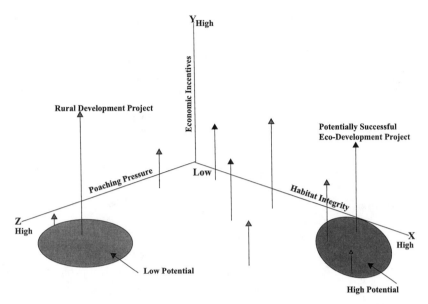

FIGURE 10.18 Model for eco-development projects targeting megafauna. This model predicts where eco-development of tiger conservation units (TCUs) would have the greatest effect on endangered species and habitats. Each vertical line represents the location in a three-dimensional space created by the intersection of three axes: habitat integrity (including resilience), poaching pressure on tigers and their prey and the ability to mitigate pressures, and economic incentives that are widely shared and powerful enough to offset threats to tigers and tiger habitat.

ability to mitigate those pressures; and strength of economic incentives to encourage nearby village communities to embrace landscape-scale conservation.

The high end of the z-axis of figure 10.18 indicates high poaching pressure and isolated fragments of habitat. Eco-development projects in these areas are essentially rural development projects and have little potential for long-term conservation. In contrast, eco-development for conservation will work best in situations that meet the conditions found on the high side of the x-axis: landscape integrity is high, poaching is low or easily mitigated with proper patrolling, and economic incentives have a good chance of promoting local guardianship of wildlife and habitats.

HABITAT INTEGRITY AND RESILIENCE. Some landscapes are more resilient than others. If the landscape consists of a habitat type such as floodplain grasslands, savannas, dry woodlands, thorn scrub, or mangroves, chances of success for an eco-development project are better. Alternatively, if the eco-development project is located in a more fragile habitat type, such as tropical moist forest, it must contain or be adjacent to a very large, un-fragmented block of habitat. Most megafauna habitats in Asia are fairly re-silient because of annual monsoons; even the driest parts of these habitats in India are likely to rebound quickly with the removal of grazing cattle (Ullas Karanth, personal communication, 1996). Thus block size and spa-tial configuration (e.g., connectivity, width of dispersal corridors, size of core areas) contribute most to habitat integrity. This variable is the least re-sponsive because large parts of the megafauna ranges are likely to remain fragmented for the foreseeable future.

POACHING PRESSURE. Because key megafauna species are so resilient across their range, they are highly responsive to the mitigation of poaching (Spitsin et al. 1987; Sunquist 1981). Tigers rebound quickly if protected and if the integrity of their habitat and prey base is intact. Rhinoceros and ele-phants rebound much more slowly than tigers but rapidly enough to show dramatic increases under protected conditions during the course of a decade.

ECONOMIC INCENTIVES. The third axis may be the most critical. For in-centives to work, they must do three things: give immediate benefit to stake-holders, involve many or most of the community, and be of sufficient mag-nitude to offset the short-term economic gains from habitat degradation and poaching. In heavily populated high-priority areas, nothing short of major revenue-sharing schemes will guarantee the persistence of rhinoc-eros, elephants, and tigers. Without these incentives, strict protection is the only effective tool for conserving megafauna. However, this approach will be limited to isolated reserves as surrounding landscapes become perma-nently altered.

Prerequisites for Adapting the Chitwan Project to Other Conservation Areas in Asia

Why is the eco-development project succeeding in Chitwan? For several reasons, the Chitwan project falls in the high potential area of the model that appears in figure 10.18. Chitwan is a very forgiving, dynamic land-scape—a floodplain habitat with high resilience where species are adapted to high levels of disturbance. Poaching pressure on rhinoceros, elephants, and tigers and prey is low, and powerful economic incentives are already in place for conserving large landscapes.

The project has been a success for other reasons. First, the virtual absence of powerful firearms reduces poaching pressure. Second, the law-abiding nature of Nepalese citizens works in favor of conservation. In any case, the Nepalese army stationed in the reserve actively discourages illegal activities. Third, the passionate commitment of a local villager, Shankar Choudhury, shows that the efforts of a single individual on one small plot of land can start a process that conserves a larger landscape. Choudhury spearheaded the on-farm forestry project on his own property in 1988 and organized the village committees to experiment with plantations and re-generation areas. Fourth, the economic incentives and the enabling legislation were strong enough to address the magnitude of the threats to wildlife and their habitats.

Lessons Learned

The experience in Chitwan and observations of other similar projects help identify some useful guidelines for locating and designing eco-development projects to meet wildlife conservation goals. Serious consideration of these recommendations could multiply the effectiveness of eco-development projects.

• Locate eco-development projects where nature is on our side. The *development* part of *eco-development* inevitably leads to a reduction or degradation of some fraction of biodiversity (temporarily or permanently). The best way to ensure the minimum loss of biodiversity is to locate eco-development projects in the most resilient habitats (e.g., mangroves, grasslands, savannas, riverine or flooded forests, and conifer forests). The biota of these ecosystems bounce back more quickly and are more accepting of the disturbance regimes created by interventions than are those in more fragile habitats (e.g., tropical moist forests, tundra, and lakes). Most megafauna species reach higher densities in more resilient landscapes, such as floodplain grasslands and savannas, than in fragile habitats like dense tropical moist forests (although important exceptions include those species restricted to rain forests).

• In more fragile habitats, species typically occur at low densities and require large areas to maintain viable populations. In such cases, the design of eco-development projects should include very large areas with an extensive core reserve. The large size of the core areas allows for mistakes or poor stewardship in the early stages of project implementation. Large areas also permit recolonization by previously exploited species populations where extraction (logging or other types of extractive measures) in the eco-development target area has not been well managed.

• Eco-development programs should never be considered as geographically isolated projects but as an integral part of a comprehensive landscape- or ecoregion-scale conservation strategy. Specifically, an eco-development area should be linked to adjacent sites with more restrictive management. Such an approach ensures that those elements of biodiversity that are eroded or lost in the project area are still conserved in the larger landscape. As an example, a project in southern Africa (Caprivi Strip, Namibia) did not want to include lions in a buffer zone because they compete with sport hunters for wild buffalo. However, the buffer zone supports so many buffalo that some wander into an adjacent park. Here they serve as prey for lions, and the lion population is well protected (Jo Tagg, personal communication, 1998).

• Eco-development projects are more likely to have a conservation effect if the immediate goal is to take the pressure off a protected area and to maintain wildlife corridors by extending buffer zones rather than to attempt to conserve all elements of biodiversity within the project area. Eco-development projects are not substitutes for strictly protected areas; they will fail if evaluated using the same criteria.

• All eco-development projects will result in a net loss of biodiversity. Be clear about the trade-offs, state them explicitly at the beginning of the project, and determine thresholds beyond which further loss is unacceptable.

• Monitor conservation effects at several levels of biodiversity: species, critical habitats, landscapes, and the ecological processes that maintain biodiversity. Tailor monitoring efforts to the type of ecosystem—for example, projects located in mangroves, estuaries, sea-grass beds, or coral reefs. In some instances, ecological processes may be far more important to monitor than species abundance or composition.

• Allow for uncertainty in the design of the eco-development project, particularly in the area of landscape management. The role of dispersal corridors — their size, extent, and condition — in the context of conservation biology has a good theoretical understanding, but little empirical data exist to guide corridor design (Beier and Noss 1998). For large mammals, corridors are likely to be the most crucial landscape elements in human-dominated landscapes. Planners should err on the side of caution by setting aside and protecting corridors larger than the minimum estimate.

• State explicitly the linkages to biodiversity conservation of each project intervention for both biological and community-based activities. The single most cost-effective means to improve the conservation effect of eco-development projects is to use the best biological insights at the design phase and throughout implementation. Ensure that a biodiversity specialist familiar with the rudiments of experimental design is involved. Local participation in monitoring is vital, but a trained biologist is essential for designing and overseeing adaptive management, evaluation of trends, and other, more technical, aspects of monitoring.

• Reinforce anecdotal accounts of the success of the project with data (maps, tables, graphs, etc.) that demonstrate the trajectory of indicators being monitored.

• Communicate important aspects of the monitoring and evaluation program to decision makers and local stakeholders through maps, posters, and powerful visuals.

The history of endangered species and habitat conservation in Chitwan demonstrates that short-term gains are achievable through strict protection, even if local residents do not share in the benefits. Over the long term, however, landscape management for area-sensitive megafauna requires partnerships with locals. To develop such partnerships, the most important step that conservationists can take is to lobby for the recycling of park revenues into local development (as a minimum for a people–nature partnership), as is now the law in Nepal. Without giving local residents tangible incentives, it will be hard to make the case for making room for megafauna in an increasingly crowded Asia.

CHAPTER 11

THE RECOVERY OF RHINOCEROS
AND OTHER LARGE ASIAN MAMMALS

Lost in the debate between proponents of in-situ and ex-situ
conservation is a simple fact: To date, the only proven method of
achieving success in rhinoceros conservation is providing good
protection of populations in effectively managed reserves.
—Nigel Leader-Williams, personal communication, 1992

THE PHANTOMLIKE JAVAN RHINOCEROS, perhaps the
rarest large mammal on Earth (figure 11.1), lives in Ujung Kulon
National Park on the island of Java; a population of fewer than ten
individuals also hangs on in Vietnam. In 1989, I tried to photograph one of these elusive animals by placing an infrared camera
trap along trails that the rhinos were thought to frequent in the
Java park. In a few days, I had amassed intimate nocturnal portraits
of rusa (deer), birds, and tree branches but no rhinoceros. Trudg-
ing around in the mud of Ujung Kulon in search of Javan rhinos ri-
valed picking off leeches in the peat swamp forests of peninsular
Malaysia while tracking Sumatran rhinos; both are an exercise in
futility. These episodes represent the practice of what I call needle-
in-a-haystack ecology. The main activity of this discipline is down-
right depressing: an extraordinary amount of field time devoted to
essentially documenting the ecological extinction of a species. Like
giant forest ox (kouprey) in Cambodia or thylacines in Tasmania,
a few Javan rhinoceros still may live in the wild, hiding during the
day, feeding secretively at night. But for all intents and purposes,

FIGURE 11.1 A Javan rhinoceros photographed by infrared camera trap in Ujung Kulon National Park, Indonesia. (Courtesy WWF Indonesia and BTNUK, Foead, Yahyo and Sumiad)

the species in question are as good as extinct because they no longer play a functional ecological role.

The rapid and catastrophic decline of Asian and African rhino populations across much of their range has prompted drastic emergency measures to prevent their extinction. Such measures include sawing off the horns of free-ranging black rhinoceros and trapping the scattered, isolated populations of Sumatran rhinos for captive-breeding facilities. I have concluded that small-scale, piecemeal attempts will not further conservation and may even do more damage. However, a combination of certain essential actions could make a tremendous difference not only for Asia's rhinos but also for the entire array of large mammals. These are strict protection of core protected areas; powerful economic incentives and innovative and supportive legislation to promote local guardianship of endangered wildlife and habitats and landscape-scale conservation; effective antipoaching units and information networks; and bold leadership to carry out essential conservation measures. These measures could include translocations, fair resettlement of villages, and equitable distribution of ecotourism dollars or other types of revenues. The Terai Arc — one of the most ambitious wildlife recovery projects to date — was designed in accordance with this comprehensive strategy and is introduced as a model for the future of large mammal conservation in Asia. The initial success of this and other efforts in economically poor and densely populated Nepal suggests that hope remains for other South Asian locations as well.

Emergency Conservation Strategies

Relocation to Habitats in Industrialized Nations

Emergency conservation strategies result from the perceived failure of traditional approaches (i.e., guns and fences) to conserve endangered mammals in their native habitat. One of the most radical proposals for conserving populations of rhinoceros and other endangered large mammals is to shift them to presumed safe locations in North America. The idea is to move rhinos, tigers, and other species now, and then reintroduce them into their native habitats when conditions are more "stable." With this purpose in mind, many Texas game ranches have become veritable designer ecosystems (Michael Soulé, personal communication, 1998), populated by ungulates from Africa and Asia intermingled with native white-tailed deer. Until a few years ago, two Asian antelope species — black buck and nilgai — were more abundant on Texas game ranches than in India or Nepal.

Why not add black or one-horned rhinoceros to the roster of Texas fauna? This approach is unlikely to achieve the desired results for many reasons. First, relocation exposes the transferees to new environmental regimes: pathogens, food plants, predators, and climate, to name a few. In addition, transport and the change in diet and conditions can cause psychological stresses. Black rhinos translocated from southern Africa to Texas did not fare well (Steve Osofsky, personal communication, 2000). Furthermore, protecting commercially valuable species is difficult everywhere. We have no evidence that rhinos would be any safer from poachers in remote parts of the United States than they would be in their native habitat. American black bears are poached extensively for their gall bladders in U.S. national parks and surrounding private lands. Finally, sufficiently large natural habitat blocks for populations of rhinoceros are unlikely to exist in the future if we do not guarantee protection of these large areas now.

Dehorning Rhinoceros

Some governments have dehorned free-ranging individuals to reduce the incentive to poach. This response was developed in desperation: poaching was rampant in the early 1990s, and resources were inadequate to protect wild black rhinoceros in large remote reserves. One advantage of dehorning is that animals remain in their native habitats and established territories. Another advantage is that in remote areas, where frequent antipoaching patrols are too costly, dehorning dissuades poachers from even searching for rhinos. Theoretically, only a portion of the entire population needs to be dehorned because, given the poachers' high failure rate, they would eventually give up hunting for rhinos with horns. The word would spread among poaching rings that the costs far outweighed the benefits of a successful poaching expedition.

Namibia and Zimbabwe carried out dehorning programs with the intent of deterring further poaching epidemics. After several years of effort, however, these governments regard dehorning negatively. The unit cost of dehorning is high—some estimates put the cost at $1,500 per individual. Compounding the high unit cost is that rhinoceros horn regenerates quickly (Berger 1994; Laurie 1978). Poachers consider even a few centimeters—the first year's growth—valuable. In many areas, poachers appear to have killed dehorned individuals out of spite, or they shot first and looked later to find that the animal was hornless. It is easy to critique a program that was experimental and lacked the luxury to dehorn rhinoceros as part of a controlled study of horn function in antipredator behavior. Although their findings are disputed by Lindeque and Erb (1995), Berger (1996) and Berger and Cunningham (1996) suggest that dehorning of female black rhinos in Namibia exposes calves to high predation risks by hyenas and lions. Clearly, the Namibia experiment was a desperate attempt to save a nucleus population until poaching could be brought under control. Even if initially successful, dehorning is only a stop-gap measure, impossible to maintain once a decimated population has recovered beyond a few dozen individuals.

Provisioning the Commodity

Another suggested solution is the creation of rhinoceros farms to produce horn commercially. Such farms have been set up to extract bile from black bear and to harvest tiger bones. Revenues generated from the legal sale could then be recycled into conservation. Rhinoceros would not be killed or subjected to painful treatment, as are the tigers and black bears. However, both the negative reaction of the international conservation community and the low fecundity of captive populations make rhinoceros farming unattractive. As with other species, horn from wild animals probably would be more highly valued than that from captive individuals. Both suggestions ignore the myths about the use of rhino horn discussed in chapter 2.

Captive Breeding of Rhinoceros and the Ex Situ/In Situ Debate

No strategy has provoked greater debate in the conservation community than captive breeding (Andau 1995; Foose, van Strien, and Khan 1995; Leader-Williams 1993; Rabinowitz 1995; Sumardja 1995). It involves the removal of small isolated populations from the wild to captive-breeding facilities. Proponents regard this strategy as a hedge against extinction in the wild. The rescue of "doomed" individuals from natural habitats slated for logging or conversion (e.g., oil palm and rubber) offers a chance for these animals to survive, breed, and help maintain the genetic vigor of the population. New understanding about management of captive ungulate populations may help increase population numbers rapidly and shorten the population bottleneck. Captive-breeding facilities provide the opportunity to explore sophisticated reproductive technologies, which some regard as the future for dwindling populations of highly endangered species. The ultimate goal of captive breeding is the reestablishment of descendants of captive populations in the wild when "suitable areas have been secured" (Foose and van Strien 1997).

Many believe that a species taken from the wild and held in captivity is no longer a true representative of that species. Captivity shields these individuals from the predators, parasites, seasonal declines in food quality and abundance, fluctuating environmental conditions, and, in the case of rhinoceros, floods that shaped their evolution over eons. Therefore, natural selection no longer operates on captive individuals (Dinerstein and Baragona 2000). This is why captive breeding should be regarded only as a final resort.

Nevertheless, the officers of the Asian Rhino Specialist Group, which is affiliated with the International Union for Conservation of Nature and Natural Resources and is charged with assessing the conservation status of rhinos and designing recovery efforts, have for decades promoted captive breeding. The debate reached a climax when the group endorsed a proposal to remove Javan rhinos from Ujung Kulon National Park, Java. The Indonesian government rejected the proposal, to the relief of most field biologists. Critics of captive breeding point out that the strategy has proved to be costly and only modestly successful with the more abundant species (black rhinoceros and white rhinoceros). They also highlighted the well-documented, remarkable recoveries of several wild populations when protected from poaching.

Most conservationists know little about a disastrous captive-breeding attempt financed by the Sumatran Rhino Trust, a consortium of Western and Asian zoo organizations. The trust spent at least $2.5 million to capture forty "doomed" Sumatran rhinos and send them to captive-breeding stations in Western and Asian zoos. Thomas Foose, one of the architects of the captive-breeding program for Sumatran rhinoceros and the lead scientific consultant to the Asian Rhino Specialist Group, and Nico van Strien, senior conservation officer, summarize the failure in the latest update of the group's action plan:

> The 1989 version of the Asian Rhino Action Plan had placed great emphasis and expectation on ex situ programs for Asian rhinoceros. However, traditional captive methods and programs have proven unsuccessful for the Sumatran rhinoceros despite investment of considerable time and effort (Foose 1996). A major part of the problem has been attributed to the unnatural conditions: e.g. diet; size and complexity of enclosures; social configuration of the sexes; and climate, including protection from excessive sunlight, especially ultraviolet.
>
> (Foose and van Strien 1997)

Recently, Foose and van Strien (1997) revealed the extent of mortality in captivity: twenty-three of the forty rhinoceros (60%) captured in the early 1980s are dead. Even worse, recruitment in captivity was zero until a calf was born at the Cincinnati Zoo in 2002. The seventeen live rhinoceros (5 males and 12 females) are scattered among ten facilities. One facility in peninsular Malaysia holds five animals, but none of the other breeding centers houses more than three. Distributing the animals in this way was supposed to maintain the genetic purity of the three putative subspecies — however, genetic analyses of the population do not support keeping them separate (Amato et al. 1995). Each zoo contributing to the trust wanted to display a pair of rhinoceros, which probably dictated their distribution. All facilities kept males separated from females or did not give them adequate space in breeding centers to engage in courtship chases — a behavior that may be necessary for successful copulation and insemination.

The attempt to obtain Javan rhinoceros, followed by the debacle with Sumatran rhino, forced a review of the position of the Asian Rhino Specialist Group (Leader-Williams 1993; Hutchins 1995). Based on disappointing results of captive breeding in the more abundant black rhino, many conservationists have come out against efforts to remove wild populations of the more rare species such as Javan and Sumatran rhino (Leader-Williams 1993). In contrast, the captive program for the greater one-horned rhinoceros has been deemed a success and supposedly "provides an important back-up for the wild populations" (Foose 1992; Foose and Reece 1996). However, supporters of this strategy ignore the fact that in every location where wild populations of rhinoceros have been protected from poaching, they have higher recruitment than captive populations (Leader-Williams 1993). No captive population can match the high rate of recovery observed in white rhino in the Umfolozi reserve or greater one-horned rhino in Chitwan and Kaziranga. Ironically, many calves born in Chitwan and two from the Chitwan regeneration area have been sent to zoos around the world to bolster the zoo population.

The relative cost of captive breeding is a clear disadvantage. Foose and van Strien (1997) provide some data on the costs involved in capture, infrastructure, salaries, and other indirect expenditures associated with captive breeding of Sumatran rhinoceros, but the true costs are hard to calculate. Leader-Williams (1993) demonstrates that black rhinoceros populations could be protected and maintained in certain areas for about $1,400 per km^2 a year. Maintaining black rhinoceros in captivity is three times more expensive. Where black rhinos overlap in range with elephants, the same sums spent per square kilometer on black rhinoceros extend to conserving wild elephants. In fact, $1,400 per km^2 buys much more than black rhinos and elephants; it includes other elements of biodiversity plus ecological processes, which cannot be conserved in foreign zoos.

The East African analysis by Leader-Williams (1993) was limited to the estimated cost of maintaining rhinoceros within existing protected areas. The Chitwan project provides an estimate of the costs of increasing the breeding area for rhinos in buffer areas. Arun Rijal (personal communication, 1998) estimates that from 1994 to 1998 regeneration of a 25-km^2 section of the buffer zone cost about $5,600 per km^2 for fencing materials (local villagers volunteered their labor). About fifty-five greater one-horned rhinos now use the regenerated areas for a one-time cost of $2,545 per individual. Furthermore, twenty calves were born in the regeneration area during the five-year period. The cost and time required for recruiting twenty calves in captive-breeding stations would be several times higher.

Earlier I estimated that the population of rhinos in the buffer zone could theoretically increase by 1,174 individuals. Extrapolating costs, the population could reach this target with an expenditure of $24.3 million, equivalent to the cost of a new major exhibit at a large zoo. Economies of scale in purchasing fencing material could probably bring the cost down considerably. The same investment could increase the number of breeding female tigers to more than fifty individuals. Moreover, the regeneration program facilitates dispersal by large mammals, increases protection of key riparian habitats, and provides important ecological services such as flood control.

We have clear evidence that recovery of populations in the wild can be faster and less costly for the medium and long term than captive breeding. Nevertheless, the divide between captive-breeding proponents and field conservationists will likely continue because of deep philosophical differences. Advocates of captive breeding tend to concentrate on the maintenance of genetic potential rather than on the ecological role that species play in the landscape. This fundamental difference helps explain why some enthusiastically support captive breeding, while conservation biologists find it objectionable in most cases. The latter group holds that true conservation of a species is impossible unless it occurs within its native habitat. A Sumatran rhinoceros that has become habituated to having its ears scratched by a keeper in a Western zoo may be a charming tourist attraction, but it is no longer an important element of Southeast Asian rain forests.

By now it is obvious where my sympathies lie. My personal experiences with rhino conservation force me to question a strategy that fails to address shrinking wildlife habitats. Planned logging concessions and agricultural conversion continue to permanently alter vast areas of productive habitat within the range of the Sumatran and Javan rhinos. Even areas left standing as forests are becoming more accessible to hunters and are subsequently poached out. We must put all our efforts into conserving large blocks of natural habitat now, or they will no longer exist in two decades. Then we will have no place to put back the "surplus" captive animals.

Approaches to Endangered Species Conservation: Visionary or Business as Usual?

Conservation efforts on behalf of endangered species — in this case, rhinoceros — fall into four categories:

1. *No action.* Doing nothing will result in extinction of rhinoceros from the wild.
2. *Business as usual.* Unfortunately, many widely accepted techniques or conservation interventions do little to halt the decline of populations, species, and habitats.
3. *Visionary moves.* Making bold moves, often requiring strong political will and courage, will significantly change the course of conservation.
4. *Quixotic ideas.* Some ideas are so far removed from the current social reality that they are difficult to imagine as having any practical value, and publicly advancing them brings ridicule. Nevertheless, changes in attitude resulting from the passage of time and circumstances may reveal the wisdom of such ideas.

My experience in the conservation field, and in Chitwan in particular, has taught me that business-as-usual practices in conserving large Asian mammals offer little braking power against the ongoing extinction crisis. However, ideas currently considered visionary or even quixotic, such as those discussed in the next section, can, over a short period of time, become reality or business as usual. Many examples show us that we should never be satisfied with the status quo and instead work toward grander goals. In the world of conservation, as in the world of politics, Berlin walls do come down.

Visionary Approaches

In 1973 the creation of a network of protected areas in Nepal was surely the most visionary of decisions. Only years later did the government recognize that this was also an extremely practical decision, even in a desperately poor country. Nepal's foreign exchange is dependent on tourists, and more than 75% of those visiting Nepal each year include one or more national parks in their itinerary.

Royal Chitwan National Park is fortunate to be the location of many visionary projects guided by bold and courageous individuals. The translocation of seventy-eight greater one-horned rhinoceros from Chitwan to the Royal Bardia National Park (figure 11.2) is a recent example. Over the objection of some government officials, Hemanta Mishra, then member secretary of the King Mahendra Trust, pushed vigorously for the return of rhinos to Bardia (Mishra and Dinerstein 1987). The first thirteen rhinoceros translocated in 1986 fared well enough to allow a second translocation of twenty-five individuals in 1990. In the second translocation, park officials and biologists captured five rhinos per day from Chitwan, loaded them into trucks, and released them in Bardia. Clearly, a field operation that was derided as unworkable by some in 1986 (when Mishra proposed it) had become reality, or business as usual, by 1990. For his willingness to bring vision to conservation against difficult odds (bureaucracy), Mishra received the Getty Prize for Conservation. Nepalese officials directed other rounds of translocations in 2000 and 2001 (chapter 5), and rhino translocations are

FIGURE 11.2 A rhinoceros captured in Chitwan awaits translocation to Bardia. Chitwan has been and will continue to be the source pool for founder populations within the historical range of greater one-horned rhinoceros. Ensuring the survival of greater one-horned rhinoceros requires the establishment of at least ten populations of at least 100 individuals. (Courtesy Broughton Coburn/WWF)

now a fundamental component of the Terai Arc recovery project (see the last section of this chapter).

In 1987 Shankar Choudhury, a forest ranger with the King Mahendra Trust, convinced local villagers, park officials, Department of Forestry officers, and a skeptical me that a very small seedling project could lead to a major program of local guardianship of endangered wildlife species and their habitats. Choudhury's hard work and conviction sustained the project through its infancy. The project evolved into the regeneration of the buffer zone, and the same idea is now a key component in the Terai Arc recovery program. The restoration of the buffer zone by using powerful economic incentives and local guardianship has quickly moved from a visionary approach to business as usual.

In 1995 Nepal's parliament enacted legislation that channeled as much as 50% of annual park revenues to local communities. Twelve years passed from proposal of the legislation until recycled revenues started to flow. But now that the legislation has become part of the everyday fabric of local development, conservation in Nepal can move forward at a much quicker pace.

A Comprehensive Strategy for Conserving Asia's Wildlife

The optimum strategy for the long-term viability of rhinoceros and other large mammals combines the following essential actions:

• *Design conservation landscapes with large core areas that conserve key biological resources and offer strict protection from poaching.* Protected areas are the cornerstone of biodiversity conservation (Noss, O'Connell, and Murphy 1997). These areas should include a core area, or an area that supports or could support a breeding population (one in which recruitment exceeds mortality) and has a high level of protection (Noss et al. 1999). Many Asian protected areas suffer from severe degradation along their boundaries, and some lack any core area undisturbed by humans. In Chitwan, breeding female rhinoceros and tigers cluster in the *Saccharum spontaneum* grasslands distributed along the edge of rivers and streams. Protection of these forage-rich grasslands is crucial to achieving the goal of expanding populations. The main reason that rhinoceros have rebounded so strongly in Chitwan is the strict protection of the breeding population in these high-density core zones. Similarly, riparian habitats are key areas for tigers as dry-season refugia and for supporting their prey base; these should also be well represented in core protection zones.

Strict protection of core areas is essential for the recovery of Javan rhinoceros in Cat Loc and Nam Cat Tien in Vietnam and in Ujung Kulon National Park in Java. Gunung Leuser, Kerinci Seblat, and Barisan Selatan National Parks in Sumatra, Indonesia, and Taman Negara in Malaysia for Sumatran rhinoceros need to take similar measures. Strict protection of core areas set the stage for creative management experiments in buffer zones. These areas are easily repopulated by megafauna when an effective core area is nearby. Moreover, use of the buffer zones by ecotourism programs means that the effective core area can be extended because of the reduction of tourist traffic inside a park.

Some rhinoceros sanctuaries in Asia are very small and have no buffer zone. Consequently, the core area equals or is the same as the conservation unit. In such parks, intensive habitat management to increase densities of target species is essential. For rhinoceros, tigers, and other area-limited species, management activities should include maintenance of year-round wallows and stream flow, prescribed burns, maintenance of short grasslands and grazing lawns, and enrichment plantings of fruit trees for large frugivorous birds and mammals. Research in the Chitwan buffer zone has shown that the application of these management techniques can increase densities. Intensive management is critical in isolated, small reserves that are too fragmented to allow dispersal.

• *Introduce powerful economic incentives, new legislation, and public awareness campaigns.* The future of Asia's megafauna rests not with biologists or managers but with local people. Populations of large Asian mammals, especially those of rhinoceros and tigers, can be considered secure only when local residents view them as being worth more alive than dead. Our work in Chitwan has shown that, given the alternative, local villagers will make decisions that promote local guardianship of wildlife habitats and endangered species. The ultimate challenge for conservationists during the next few decades is to promote the extension of public stewardship of natural resources to more areas and at grander scales.

A common misunderstanding is that local villagers are indifferent to the degradation of nearby habitats. The value of forest cover in riparian corridors became quite obvious to communities in the Chitwan Valley during the 1993 monsoon floods. Where riparian forests still stood, many houses were saved. Where the buffer-zone forests had been cut down, whole villages washed away. Beyond flood control, managed forests also provide access to fodder and firewood. Conservation of riparian forests, grasslands, and wetland habitats is a first major step to regenerating large landscapes in South Asia for the benefits of human populations and wildlife.

Ecotourism can become a powerful tool for convincing communities to respect the sanctity of core areas and to create effective buffer zones and corridors if the flow of revenues to local groups is rapid and substantial. Ecotourism is one of the fastest-growing industries in the world, and conservationists should use it as financial leverage whenever possible. Indigenous ecotourism is growing more rapidly than foreign tourism in Chitwan; the Nepalese are now the leading visiting nationality, followed by tourists from neighboring states in India. Inconceivable five years ago, South Asian tourism is important because it partially offsets the vagaries of international ecotourism. The middle class in India — a group awakening to the pleasure of visiting nature reserves in their own country and in neighboring Nepal — is larger than the entire population of the United States. Indian conservationists should push hard for adoption of legislation similar to Nepal's to allow for revenue from park entry fees and concessions to be plowed back into local development. Once incentives are in place, planned and unexpected developments will blossom more quickly and more effectively. This is not to suggest that reserves with fewer immediate or long-term prospects for ecotourism are a lost cause. Rather, we should take advantage of those reserves that have the infrastructure in place to promote ecotourism and find other mechanisms for reserves too remote to attract many tourists.

Public awareness is an essential part of promoting local guardianship. Nepalese living around Chitwan, particularly the younger generations, are aware of the unique status of greater one-horned rhinoceros and of the global importance of the park. In Chitwan extensive outreach by the Department of National Parks, the King Mahendra Trust, and others have increased local appreciation of the park's wildlife. Nationally, television and radio shows and nature documentaries filmed in Chitwan have encouraged wider support.

• *Organize effective antipoaching information networks and an antipoaching unit.* Even the most powerful economic incentives will not stop exploitation of species such as rhinos, if they command a high value on the commercial market. Currently, however, protection measures for rhinoceros populations are underfunded throughout Asia (Rabinowitz 1995). The use of incentives to conserve species, augmented by effective antipoaching information networks and antipoaching units, forms the backbone of a sound recovery strategy.

The invisible network of local people who form the antipoaching information network is as important as the visible presence of armed guards. Most rhino poaching during the 1990s occurred in the buffer zones, perpetrated by outsiders who are easily identified in surrounding villages. Under the antipoaching program in Chitwan, informing on poachers is more profitable than poaching a tiger or rhinoceros. The activities of informers can also make patrolling far more strategic and cost effective. Unfortunately, effective networks are more the exception than the rule across South Asia. Conservation donors and government agencies should give priority to funding these units, for these networks, combined with other incentives, have proved to be effective at quickly reversing the trajectory of declining populations.

Ecotourism programs also contribute to antipoaching efforts. Tourist lodges in Chitwan are dispersed throughout rhino habitat. Hotel owners, managers, and nature guides have a strong incentive to show rhinos living near their viewing areas. They are quick to alert park staff about the movements of suspicious individuals in their areas or of a poaching incident.

• *Identify bold leadership to rally the political will to carry out essential measures, such as translocations, redistribution of ecotourism revenues, and fair resettlement, and to promote landscape-scale conservation.* The conservation of Asia's megafauna is not for the fainthearted. Advocates and critics are everywhere, offering advice to officials in range states about how to save their indigenous species. Foreign groups and individuals play a valuable role in raising awareness and money to help local conservation efforts, but grassroots efforts to save wildlife and habitats have no peer. Conserving Asian megafauna does not require a majority vote but the passion and commitment of a few charismatic individuals. No individual in Asia did more to save wild tigers than Prime Minister Indira Gandhi. The royal family of

Nepal has propelled Nepalese conservation to the forefront and provided a basis for long-term efforts. Now, younger Asian conservationists must step forward to be a voice for the preservation of their natural heritage. The courage of government wildlife officials and local leaders, not the enthusiasm of foreigners, will determine the fate of Asia's megafauna. Local leadership is critical for the following types of action:

Translocation of rhinoceros. At least five areas that once contained large populations of greater one-horned rhinos — Corbett, Manas, Bardia, Dudhwa, and Jaldhapara — could be or are being repopulated with individuals from Chitwan and Kaziranga (see figure 3.2). Some smaller reserves, such as Sukla Phanta in western Nepal and Dudhwa in northern India, could support small breeding populations if managed intensively as part of a metapopulation. Sumatran rhinoceros populations that are widely scattered and difficult to protect should be translocated to form a few larger populations in several well-patrolled areas in Malaysia and Indonesia. Very soon, Indonesian wildlife officials must translocate a portion of the Javan rhino population from Ujung Kulon to Barisan Selatan National Park in southern Sumatra or another protected area.

The reluctance to translocate the Southeast Asian species of rhinos stems from the fears of wildlife officials about the effect of capture and translocation. However, these fears can also be an excuse for inaction. Our research program in Nepal shows that greater one-horned rhinos can be captured and transported safely using standard immobilization techniques (Dinerstein, Shrestha, and Mishra 1990). The recovery of Javan and Sumatran rhino populations clearly requires bold leadership for implementing essential translocation programs. Similar courageous acts from Assamese wildlife officials are essential to overcome local reluctance for translocation of rhinos to other reserves in Assam, West Bengal, Bihar, and Uttar Pradesh.

Asian and African wildlife officials and biologists need to meet to exchange their experiences in translocation. For example, when animals are transferred in several waves separated by several years, data from South Africa demonstrate the importance of moving adult rather than subadult animals in the subsequent translocations. Reintroducing subadults to areas where a rhinoceros population is already established can result in increased mortality (Jacques Flamand, personal communication, 1998).

Those who wish to maintain the purity of the putative subspecies of Southeast Asian rhinos sometimes block translocation programs. But holding out for the ideal source populations for translocation appears to be a case of the perfect's being the enemy of the good. Modern conservation biology may need shortcuts to achieve goals, and they may not entirely satisfy the needs of other disciplines, such as ungulate taxonomy and population genetics.

Redistribution of park revenues to local communities. In almost every country where tourists visit parks and protected areas, all or a substantial portion of the revenues generated goes to the central government. Nepalese government officials offered a sound argument for this arrangement: "Nepal's foreign tourists are not going to spend a rupee to visit a leper colony or a water supply project. The national parks are a source of revenue for the entire nation to improve the lives of its people." While this argument is certainly valid, it means little to the rural poor who live near or inside reserves that attract foreign currency. Now, through landmark legislation, the government of Nepal has found that tying the welfare of people living around parks to the parks themselves promotes the long-term survival of both. Moreover, the wide coverage of the park system means that, in theory, many villagers in remote areas could benefit from the redistribution plans currently in effect around Chitwan and Bardia. Some programs may reach those in buffer zones faster than will federal programs run from Kathmandu or district headquarters, where administrative delays are common. Rapid reinvestment of proceeds can strengthen the association between conservation and social welfare.

Changing national legislation to recycle park revenues is the single most powerful intervention that government officials and conservationists in other countries can implement to improve protection of endangered species. As chapter 9 shows, nature tourism, as practiced in Nepal, is an exploitative industry. Without this legislation, owners of tourist lodges would continue to benefit disproportionately. Without legislation to redirect a portion of profits to local development, villagers around Chitwan would have no incentive to support conservation.

Resettlement or land transfer. Some village areas occupy isolated enclaves within reserves. These groups have little or no access to markets, educational and job opportunities, or better health care because their isolation is the result of large flooding rivers, dense jungle, mountain ranges, and other barriers. Large and small herbivores — wild boar, deer, rhinos, monkeys, parakeets, and wild elephant — often threaten their crops (Milton and Binney 1980). In Chitwan, the 20-km^2 area known as Padampur is one such enclave.

Located on the southern bank of the Rapti River, the population of 14,000 individuals is largely cut off from the town and city centers of Chitwan District during the monsoon season. In some years, monsoon floods threaten to inundate all the croplands, as was the case in Padampur in 1993, prompting a wave of poor, landless refugees, seeking shelter and farmland elsewhere.

Not surprisingly, the people of Padampur agreed to move to an area near the Barandabar forest that had been set aside by the government for flood victims. By the summer of 1999, the land transfer was 40% complete. The government arranged to compensate farmers with a maximum of about 5 ha, a supplement in cash, and transportation of their belongings and disassembled houses to the new site. Villagers expressed some initial discomfort, especially concerning the distance to obtain water. However, a poll showed that overall satisfaction was much higher in Barandabar than in the old location.

Many people, including human rights activists and anthropologists, typically react strongly to any scheme that involves resettlement. For resettlement to be a valuable tool for landscape management and poverty alleviation, it must be creatively and equitably structured. First and foremost, resettled villagers should receive more amenities in their new location than they had in their old location. Bold leadership and vision on the part of local officials and village leaders are equally important.

The benefits to wildlife from resettlement can be considerable. The conversion of Padampur to wildlife habitat (figure 11.3) will likely expand the rhinoceros and tiger population in the park by at least 5%. Threatened grassland bird species, such as the Bengal florican, have already begun to use the recently vacated village areas. They could also be important for the experimental management of grasslands to benefit tiger prey species.

Voluntary land transfer in private lands surrounding megafauna reserves where ecotourism is popular is likely to become more common. In Chitwan some households along the park boundary area in the Bagmara area have sold their land to speculators from Kathmandu who see the park boundary as having great potential for ecotourism. The value of a hectare along the park boundary (and in the officially gazetted buffer zone) near Sauraha skyrocketed to $30,000 in 1998. Consequently, villagers are now selling their land for huge profits and buying much larger parcels for farming in areas far away from the park boundary. Many of these people are subsistence farmers, so they increase their wealth dramatically instead of staying where farming is difficult because wildlife damage crops. The gov-

FIGURE 11.3 Resettlement is an essential tool for restoring landscapes for megafauna, particularly in areas that are enclaves within protected areas such as Padampur in Royal Chitwan National Park. Major flooding forced the evacuation of Padampur in 1996 to an area far from the ravages of monsoon floods. By March 1998, this section of Padampur had reverted to natural habitat and was used by rhinoceros, cervids, and the endangered Bengal florican. (Photo by Eric Dinerstein)

ernment should require land speculators who purchase buffer-zone farms to zone part of this property for regeneration. This would help resolve another problem for landscape conservation: the mitigation of "hard edges" where villages directly border park boundaries without a buffer zone.

At present, the Nepalese government is financing the entire Padampur resettlement effort on its own. Major foreign donors, a potential source of funding, are reluctant to participate in resettlement projects, fearing recrimination. Major donor agencies need to step in where land transfer is in the best interests of people and wildlife.

Support for Landscape-scale Conservation: The Terai Arc

What began in 1987 as a native tree nursery rapidly evolved into a buffer-zone restoration program by 1990. The time was ripe to promote a more ambitious idea: we could use the existing forest corridors to link the eleven reserves of the Terai zone, from the Parsa Wildlife Reserve to the east of Chitwan to Rajaji and Corbett National Parks at the western edge of the Terai (figure 11.4). We called this necklace of reserves the Terai Arc. The goal is simple and yet extremely challenging: to manage various populations of tiger, rhinoceros, and elephant as part of a larger metapopulation. A key part of the plan is the incorporation of dispersal corridors that facilitate exchange among the different core reserves. None of the existing reserves within the Terai Arc was larger than 1,000 km², and none contained more than sixty breeding adult tigers. By linking reserves and maintaining gene flow, we could create a much larger effective population size for tigers. The enormity of this restoration program seemed quixotic at the time, but in biological terms it is entirely sound. This is the scale of intervention needed to maintain the viability of large mammal populations, particularly those that are poor dispersers.

Empirical data from Indian reserves showed that, despite their large size, swimming ability, and propensity to move 20 km in a single night, tigers are averse to crossing gaps in natural habitat that are wider than 5 km (Ullas Karanth, personal communication, 1998). In contrast, rhinoceros and elephants will cross agricultural fields at night or follow watercourses to navigate through settled areas. We therefore designed the Terai Arc program around the dispersal requirements of tigers and mapped the bottlenecks within corridors — gaps in natural habitat wider than 5 km that required some degree of restoration. We have identified eight serious bottlenecks that must be bridged by 2010 (Joshi, Dinerstein, and Smith 2002). We plan to eliminate two or three of the most serious bottlenecks by 2004.

Aside from eliminating bottlenecks, the Terai Arc program contains other important elements. These factors include the successful restoration of two new rhinoceros populations of more than 100 individuals each at Bardia National Park and the Sukla Phanta Wildlife Reserve; the establishment of a trust fund to endow the corridor restoration program for the long term; and the commitment of the Indian and Nepalese governments to collaborate on transboundary conservation. Fortunately, sufficient forest remains between reserves to consider a restoration program on this scale. We expect to apply many of the techniques that encourage local guardianship and management of corridors and buffer zones to the Terai Arc as a whole. We may also experiment with new types of incentives, such as conservation performance payments (Ferraro 2001), as a substitute for eco-development

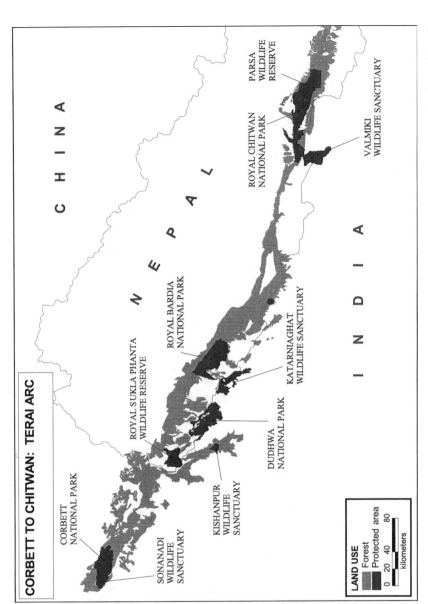

FIGURE 11.4 The Terai Arc is a landscape-scale program designed to maintain or restore connections among the eleven national parks and wildlife reserves found in the Terai zone of southern Nepal and northern India.

projects or as a supplement. Conservation performance payments provide an annual fee to local groups that have tenure to lands valuable for biodiversity, such as wildlife corridors, to maintain the forest cover or natural habitat on a per-hectare basis. This approach is now being tested in a variety of settings and offers promising opportunities. In Costa Rica, farmers and ranchers are more than willing to maintain forest cover on their properties for a compensation rate of about $35 per hectare a year. In fact, the willingness to participate outpaces the money available for compensation. In the case of the Terai Arc, we could offer local communities a graduated scale of incentives based on performance; performance evaluations would include not only the maintenance of forest cover but also the documented use of the corridors by the target wildlife species (Joshi, Dinerstein, and Smith 2002).

In November 2000, the prime minister of Nepal officially endorsed the Terai Arc, as have the minister of forests and the directors general of the Department of Forestry and the Department of National Parks and Wildlife Conservation. Many consider it to be an important crucible for experimenting with landscape-scale conservation in a developing country. First, the Terai Arc cuts a swath across an extremely poor and densely populated part of the globe. Second, much of the highly fertile land in the Terai Arc has already been converted to agriculture; remaining habitat is very productive and in great demand. Third, the wildlife species that are the targets of the corridor restoration program are large, occasionally dangerous to human life, and often destroy crops. If we can achieve success at a landscape-scale despite all these constraints, others will be encouraged to replicate the approach elsewhere. Attaining this goal will not be possible without the cooperation and leadership of government planners and donor agencies. It also requires the full participation of local people because of the critical role they play as the principle stewards of the community forests that will form the basis of wildlife corridors and buffer zones.

Where will the money come from to support a strategy to conserve Asia's megafauna? Some of the most important conservation initiatives are within the budget of every Asian country. Enacting new legislation to promote the recycling of park revenues and local guardianship, authorizing the formation of antipoaching information networks, or involving local people in designing conservation landscapes is not costly. Other interventions — translocations, the regeneration of buffer zones and corridors, resettlement, and the recurrent costs of park protection — can be expensive. First, we can put the costs of large mammal conservation in perspective. Leader-Williams (1993) equates the budget of a single large zoo in the United States in 1990 ($70 million) with the entire field protection budget of the countries of sub-Saharan Africa in 1980 ($75 million). Thus the annual maintenance budget of a major new zoo exhibit is similar to that of the operating budget of a national park in an Asian country. No one would suggest that zoos stop spending money at home. Rather, these comparisons provide some perspective on how much more in situ efforts could achieve for conservation of endangered species and populations.

I propose the following framework to secure a future for Asian megafauna:

1. Identify the most important conservation landscapes and invest in their protection. These landscapes should be of global importance for biodiversity conservation and include broad representation of species assemblages in a wide array of habitats. Tiger landscapes have already been identified (Dinerstein et al. 1997). In collaboration with the Wildlife Conservation Society, the World Wildlife Fund–United States identified 159 tiger conservation units — blocks of natural habitat where tigers occur or are likely to occur — in representative habitats across the range. Of these, we designated 25 as of highest priority for conservation action (level 1) and 24 others as of high priority (level 2) (figure 11.5). Similarly, Sukumar (1999) identified twenty populations of Asian elephants with the best chance of long-term persistence (figure 11.6). Figure 11.7 represents an overlay of these core landscapes with the World Wildlife Fund's Global 200 priority areas for biodiversity conservation (Olson and Dinerstein 1998). It shows that the Global 200 contains all the important landscapes identified for large mammal conservation, except the dry zone of Sri Lanka, which represents important elephant habitat (table 11.1). It also shows that the Global 200 ecoregions cover the range of habitat types in which large mammals are found. Therefore, in this part of the world the core landscapes for large mammal populations can serve as an umbrella for the conservation of many of the most biologically rich areas of Asia west of Wallace's Line. (Most of Asia's megafauna is limited to west of Wallace's Line.) Further refining this map would be possible by adding distributional data for primates, bears, and wild cattle. However, the landscapes already selected in the overlay presented here probably capture some of the most important areas for these other large mammal species.

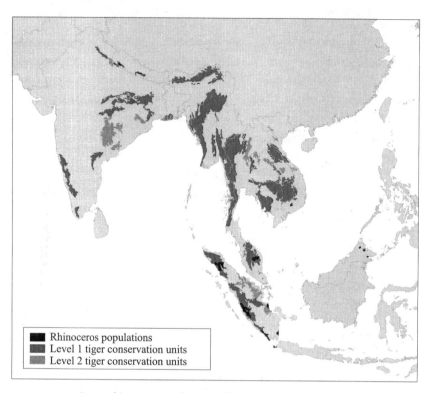

Rhinoceros populations
Level 1 tiger conservation units
Level 2 tiger conservation units

FIGURE 11.5 Extant rhinoceros populations and highest priority areas for tiger conservation in Asia. (Adapted from Dinerstein et al. 1997)

2. Identify the key activities needed to maintain and enhance large mammal populations in these areas. Much of this work has already been done through regional and national analyses.

3. Conduct a financial gap analysis to determine the extent to which donations match need and absorptive capacity. A good model already exists for Latin America and the Caribbean (Castro and Locker 2000).

4. Enlist the multilateral and bilateral funding agencies, international conservation organizations, foundations, individual philanthropists, international zoo community, and national governments to finance large-scale conservation for ten years. This would perpetuate new incentive-based approaches to conserving landscapes. Options include trust funds, carbon sequestration offsets, legislation to recycle park revenues, community-based

FIGURE 11.6 Extant rhinoceros populations and highest priority areas for wild elephant conservation in Asia. (Adapted from Sukumar 1999:47)

Rhinoceros population
Elephant priority area

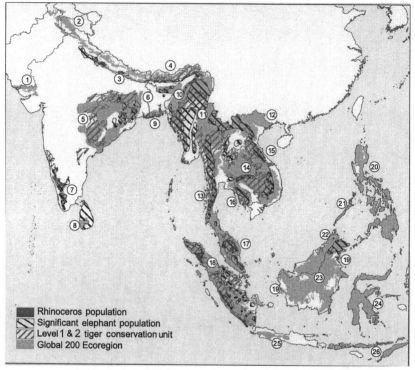

FIGURE 11.7   Overlay analysis of high-priority areas for conservation of Asian rhinoceros, elephants, and tigers with Global 200 ecoregions. The numbers correspond to the Global 200 ecoregions in table 11.1, which lists the most important landscapes for megafauna conservation within each ecoregion. Virtually all the important landscapes for large mammal conservation occur within Global 200 ecoregions. (Courtesy Meghan McKnight, Conservation Science Program, World Wildlife Fund–United States)

ecotourism programs, a bed tax, and an entry tax to benefit community-based activities in buffer zones. In addition, foreign agencies that fund development projects should require that some portion of the funding be dedicated to enhancing landscape-scale conservation, community-managed forest concessions, and other means of providing a local economic stake in endangered species and habitat conservation. Where endangered species are targets of zoo collections, the fees paid by the zoos might be deposited in trust funds partly managed by local people. For example, investing in a trust fund the money paid by zoos to capture two rhinoceros calves a year might finance in perpetuity the restoration of degraded buffer zones, antipoaching efforts, and local development programs around Chitwan.

TABLE 11.1. IMPORTANT CONSERVATION LANDSCAPES FOR LARGE MAMMAL POPULATIONS WITHIN THE GLOBAL 200 ECOREGIONS

Number	Global 200 Ecoregion	Features of conservation landscapes identified as priority areas for large mammal populations
1	Rann of Kutch flooded grasslands	Home of last large population of Asian wild ass
2	Western Himalayan temperate forests	Home of healthy populations of a diverse assemblage of large carnivores and montane ungulates
3	Terai-Duar savannas and grasslands	Levels of large ungulate biomass that rival those of East African parks; important landscapes include Chitwan-Parsa-Valmiki, level 1 TCU*; Bardia-Banke, level 1 TCU; Rajaji-Corbett, level 1 TCU; potential for important elephant corridor between Corbett and Bardia; part of Manas-Namdapha, level 1 TCU; Buxa-Manas-Bhutan elephant population
4	Eastern Himalayan broadleaf and conifer forests	Arunachal-Assam–North Bank elephant population; Garo Hills elephant population; Manas-Namdapha, level 1 TCU, and isolated populations of wild water buffalo in foothills
5	Eastern Deccan plateau moist forests	Kanha-Pench, level 1 TCU; Kanha-Indravati corridor; Similipal-Kotgarh, level 1 TCU; Orissa–central India elephant population
6	Chhota Nagpur dry forests	Bagdara-Hazaribagh, level 1 TCU
7	Southwestern Ghats moist forest	Levels of large ungulate biomass that rival those of East African parks; largest population of elephants in Asia (Nilgiri–Eastern Ghats, Anamalais-Nelliampathis, Periyar-Madurai); Dandeli-Bandipur, level 1 TCU; Periyar-Kalakad, level 1 TCU
8	Sri Lankan moist forest	Endemic flora and fauna
9	Sundarbans mangroves	Sundarbans, level 1 TCU—only potentially significant population of tigers in mangrove habitats
10	Naga-Manapuri–Chin Hills moist forests	Important landscapes include Kaziranga-Meghalaya TCU; Arakon Yoma, level 1 TCU; Kaziranga–Karbi Anglong priority elephant population; endangered large herbivore populations such as swamp deer
11	Northern Indochina subtropical moist forests	Remnant elephant population in Yunnan–north Laos; uncertain status of tiger populations
12	Southeast China–Hainan moist forests	Endemic plants and invertebrates but most large mammals have been extirpated from southeast China
13	Kayah-Karen/Tenasserim moist forests	Huay Kha Khaeng–Thung Yai Naresuan, level 1 TCU

TABLE CONTINUED

TABLE 11.1 CONTINUED.

Number	Global 200 Ecoregion	Features of conservation landscapes identified as priority areas for large mammal populations
14	Indochina dry forests	Largest tract of dry forest remaining in Indochina; Virachey–Xe Piane–Yok Don, level 1 TCU; large populations of gaur and possibly kouprey; remnant Javan rhino population in Cat Loc, Vietnam
15	Annamite Range moist forests	Hot spot of new discoveries of Asian mammals; remnant population of Asian elephants; Nam Theun Nakai–Vu Quang, level 1 TCU
16	Cardamom Mountains moist forests	Phnom Bokor–Aural, level 1 TCU; largely unexplored area that may hold new species of mammals similar to the Annamites; important elephant populations
17	Peninsular Malaysia lowland and montane forests	Large elephant population for Southeast Asia; Taman Negara–Belum-Halabala, level 1 TCU; stronghold of Sumatran rhinoceros in mainland Asia (Taman Negara National Park, Malaysia)
18	Sumatran Islands lowland and montane forests	One of last strongholds for Sumatran rhinoceros; preferred translocation sites for Javan rhinoceros; Gunung Leuser–Lingga Isaq, level 1 TCU; Kerinci–Seblat Seberida, level 1 TCU; Bukit Barisan Selatan–Bukit Hitam, level I TCU; Way Kambas, level 2 TCU; significant elephant populations in Aceh, Riau, and Lampung–south Sumatra
19	Greater Sundas mangroves	Important habitat for proboscis monkeys and other wildlife
20	Philippines moist forests	No native large mammal populations except endemic wild water buffalo (tamarau on Mindoro Island); elevated to Global 200 because of high endemism of plants and animals
21	Palawan moist forests	No native large mammal populations; elevated to Global 200 because of high endemism of plants and animals
22	Kinabalu montane shrublands	No native large mammal populations; elevated to Global 200 because of high endemism of plants and animals
23	Borneo lowland and montane forests	Only Asiatic elephant population (Sabah) and significant population of Sumatran rhinoceros (Sabah lowlands)
24	Sulawesi moist forests	Home of the babirusa and lowland and mountain anoa; highest bird and mammal endemism in Asia
25	Western Java montane forests	Abut Ujung Kulon National Park, home to largest population of Javan rhinoceros
26	Nusa Tenggara dry forests	Include islands of Komodo and Flores, home to Komodo dragons

* TCU, tiger conservation unit.

The collapse of several Asian economies in early 1998 suggests that it may be some time before they are able to allocate resources to conservation. The initiatives I have listed here buy time for Asia's megafauna until a strong conservation ethic to preserve species and their habitats evolves. Thus the reality is that citizens of wealthy nations who desire a world with large mammals must be willing to share the cost of conservation everywhere.

Is time running out for the conservation of Asia's large mammals? In a period of overall decline, I see rays of hope that will allow most species to persist in at least a fraction of their original range. Much of my optimism stems from field experiences in Royal Chitwan National Park, which in 1998 celebrated its silver jubilee. In 1991 it was inconceivable that Chitwan's wildlife populations and habitats would be in better condition in 2001. This experience demonstrates that, given adequate protection from poaching and provided with suitable habitat, even some of the largest and slowest-breeding mammals can recover quickly from episodes of near extinction.

Beyond Chitwan, parks throughout the Indian subcontinent offer hope. Large mammals are still holding on in many South Asian reserves, even in the face of poaching pressure and degradation of national parks. From Sri Lanka to India to Nepal, hunting reserves once set aside by maharajas or by colonial rulers became national parks. These parks still maintain a visible, if not vibrant, array of megafauna. The millennium could mark a new era of restoration for protected areas on the Indian subcontinent. Expanding and linking reserves (as the government of Bhutan has done), as well as translocating megaherbivores, are all possible, provided the leadership exists to push the programs forward.

The course of conservation, particularly in Thailand, Laos, Cambodia, and Vietnam, is quite different from what I have experienced on the Indian subcontinent. In Indochina wildlife reserves exist in name only; intense poaching has decimated the large mammal populations of this region. One is lucky to encounter tracks of large vertebrates in protected areas, let alone be blessed with an actual sighting. Why is this so? In stark contrast to the Indian subcontinent, Southeast Asia has no tradition of strict protection within nature reserves. But without strict protection, large mammals continue to disappear. With just a modicum of effective enforcement, these species could recover rapidly. We can only hope that conservation leaders in Indochina will take action before it is too late.

Some of the moments I cherish most came while observing large animals moving with grace and dignity in full freedom from human encroachment within national parks. Lately, I am thrilled to observe female rhinos with young calves, mugger crocodiles, and tigers in a buffer-zone area that, until recently, consisted of degraded scrub jungles and grasslands grazed to golf-course conditions. Now we walk through tall forests dominated by native *Albizia* trees, keeping a close eye out for a rhino, tiger, or leopard. We have accomplished what Michael Soulé and Reed Noss call "rewilding"—returning the land to the creatures that once flourished here. I hope that rewilding becomes the mantra of the next generation of Asian conservationists, as it takes root in other regions of the world.

As we embark on a new century, we are surrounded by examples of our own ingenuity and dominance as a species. We are sophisticated enough to capture a rhino or a tiger, fit it with a satellite telemetry collar, and let schoolchildren from around the world track its daily movements on the Internet. But a Web site cannot convey what it is like to watch a rhinoceros move through the grasslands of Chitwan, hear a tiger roar in the night, or listen to elephants rumble to one another in a subsonic language barely perceptible to our ears. The presence of large, potentially dangerous mammals connects us to something deep and primal and teaches us humility in a way that is unique and precious. Wild places where species leave footprints larger than our own must be part of the legacy we bequeath to future generations.

METHODS

Collection of Biological Data

Measurements

Between 1984 and 1988, I measured forty-eight greater one-horned rhinoceros, representing about 13% of the Chitwan population and 4% of all free-ranging individuals at the time. The sample included thirty-six adults (20 males, 16 females), six subadults (2 males, 4 females), and six calves (2 males, 4 females). We collected data on twenty-two mensural and sixteen nonmensural characteristics of immobilized individuals. We attached radio transmitters to facilitate study. We also measured individuals immobilized for translocation to other reserves or to treat wounds. Fights between breeding males resulted in early signal failure of the radio collars; replacement of transmitters gave us the opportunity to reexamine and remeasure seven adult males and two adult females. We also measured the carcasses of four adult males, one subadult male, and two calves that were less than a year old.

We made ten measurements with cloth tape: total length of body, length of tail, length of head and body, maximum skull circumference, chest circumference, neck circumference behind head, neck circumference in front of shoulder, shoulder height, distance between junction of posterior cross-skin fold and anterior edge of anal skin fold, and horn circumference at base. We used tree calipers to take thirteen measurements: length of hind foot to base of middle hoof, length of hind foot to tip of middle hoof, length of ear, length of head, cranial breadth, width behind head, width in front of shoulder, shoulder width, width across anterior cross-skin fold, width across posterior cross-skin fold, length of horn, length of lower left outer incisor, and length of lower right outer incisor. We were able to measure shoulder height and chest circumference precisely on translocated animals, which could be maneuvered from sternal into lateral recumbency during handling, and on carcasses. Weighing free-ranging captured individuals was not possible during this study.

Differentiating the degree of dimorphism throughout the life span of greater one-horned rhinoceros required classifying carcasses or anesthetized individuals into discrete age classes. However, data are lacking on known-age, captive, and wild greater one-horned rhinoceros for comparison. For adults (animals older than 6), we used tooth wear to classify adults into age categories. From studies of known-age black rhinoceros, Hitchins (1978) and Goddard (1970) identified eleven discrete age classes for adults based on cementum lines and eruption and attrition of dentition. Studies of white rhinoceros showed good correlation between known-age animals and cementum lines (Hillman-Smith et al. 1986), and researchers identified sixteen tooth-wear categories with eight classes of adults. I was not able to examine cementum lines in M1 teeth of known-age adults, so, to establish three adult age classes, I modified the technique of Hillman-Smith et al. (1986) for creating age criteria based on attrition of the molars. I used the presence of sharp ridges on the lower molars to characterize the young adult class, a relative flattening of the ridges of lower molars to typify intermediate-age adults, and well-formed depressions on the occlusal surface to identify old adults.

Observations of captive animals in zoos revealed that old adults are easily distinguished from younger animals by size and by the accumulation of several nonmensural characteristics. Thus I linked data on dentition with other mensural and nonmensural characteristics to classify adults as young, intermediate-age, or old. Mensural characteristics included horn growth and size, body size, and length of mandibular incisors. Nonmensural data included wrinkles and development of secondary skin folds, erosion of tissue around the horn, and, for females, number of calves raised. While establishing criteria for adult age classes, I assumed that wild animals should not differ from zoo animals in accumulation of physical characters associated with advanced age (horn wear, wrinkles, and added skin folds) but perhaps in magnitude. Also, I noted that horns grow at the rate of 2 cm per year in subadults and young adults, and I assumed that horns continue to grow at this rate until past middle age. I used other characteristics to estimate ages of old animals with broken horns.

To test for differences in mensural data between age and sex classes, I used a Mann-Whitney U test. To test for differences in nonmensural characteristics among sex and age classes, I used chi-square analysis. I also analyzed variation in nonmensural characteristics for animals observed but not anesthetized during this study; I included animals only if I had a complete set of photographs (front, rear, left side, right side) and detailed descriptions.

Estimation of Abundance

I identified four distinct subpopulations that had developed because of isolation by physical barriers (rivers and low mountains) or by ecological boundaries (extensive tracts of *sal* forest or cultivation): the Sauraha, the West, the Bandarjhola-Narayani River, and the South Botesimra. The Sauraha was the largest subpopulation (Laurie 1978) and the most intensively surveyed. It spanned the grasslands east of Kagendramali to the edge of the *sal* forest near Kasara in the west. Laurie (1978) separated the Kagendramali and Sauraha subpopulations, but I combined them after discovering frequent movement of animals between the two areas.

Nearly 12 km of unbroken *sal* forest separated the Sauraha population from the West; Laurie (1978) regarded this as a partial migratory barrier limiting extensive movements between areas. Most animals in the West population were concentrated within a 3-km radius of the confluence of the Reu, Rapti, and Narayani Rivers (see figure 5.2). Several kilometers of agricultural land and the Narayani River separated the rhinoceros population on Bandarjhola Island and along the Narayani River floodplain from the West population. Extensive *sal* forest and a low mountain range separated the South population from the Sauraha population, and extensive *sal* forest, a highway, and cultivation separated other populations from the Sauraha population.

I subdivided the Sauraha population into census blocks (see figure 5.2, inset) to examine habitat–density relationships. I used aerial photos to map the area of each block covered by the important forage grass *Saccharum spontaneum* and riverine forest, the two habitats that rhinoceros use most frequently (Laurie 1978). I collected most census data during the hot-dry season (February–May) after the tall-grass layer had been burned off, when visibility was at a maximum, and when rhinoceros grazed for long periods in the open. After, May grass obscured newborn calves, and identifying and sexing individuals was difficult.

I used photographs and sketches to identify individuals. Irregularities in skin folds and ear cuts provided the most striking and unambiguous characteristics for rapid field identification (chapter 4). The registry provides the unique characteristics of all individuals.

Between 1984 and 1988, I visited 95% of the park and adjacent forests where rhinoceros occur. Within each population, I made my observations, identification, photographs, and registration of rhinoceros from atop domesticated elephants trained for this task. I conducted the searches in the morning and late afternoon, when animals were most active. Habituation of many individuals to elephants increased my ability to obtain detailed photographs. When individuals were less cooperative, three assistants (also atop elephants) and I surrounded them and herded them toward the elephant supporting the photographer.

I used a pilot study on eight radio-collared animals (2 males and 6 females with calves) in the Icharni block to estimate the minimal time required to find and to identify every individual in our study blocks. On five different days, I determined the number of radio-collared individuals that escaped visual detection during each of two searches. We needed about sixteen hours of search time with five elephants (80 elephant-search hours) to find all thirty-nine animals in the 3.2-km^2 area. I made estimates of search time in the other blocks in the Sauraha population by projecting the Icharni results on an area basis ($80 \div 3.2 = 25$ elephant-search hr/km^2). In actuality, we spent far more time between 1985 and 1988 surveying blocks containing high population densities in the Sauraha population than what I had figured as the minimum estimate because we did census work and photo registration while doing other research.

I conducted censuses annually and adjusted population sizes to account for animals missed from one year to the next. Discovery of new animals, other than calves less than one year old, dropped from 10% to less than 1% from the first census year to the fourth year.

I did a census of only the West and Bandarjhola-Narayani populations in 1986, so I used my estimate for population growth between 1986 and 1988 in the Sauraha area to project numbers in 1988 in these areas. I also corrected the West and the Bandarjhola-Narayani estimates for animals that we likely missed; fresh dung and tracks in a few areas indicated the presence of animals that were not registered in either population. From data on the Sauraha population, I estimated that the 1986 census missed 7% of subadults and adults, which were subsequently added to the census in the following two years (1987–1988).

We added three calves to the 1988 cohort, as we missed three births per year (on average) that were subsequently registered in the following census year. This oversight was the result of births that occurred after the last month of census work for the year. We subtracted fifteen animals from the subadult cohort registered by 1988. We had registered these subadults as four-year-old calves while they were still with their mothers, but we most likely had re-registered them as subadults in subsequent years. Laurie (1978) also adjusted his total estimate to account for this problem.

The translocation program transferred twenty-four individuals from Royal Chitwan National Park between 1986 and 1988. Two other reserves received seventeen adults and subadults, and seven calves went to zoos. The twenty-four translocated animals came from the Sauraha population, as did another ten calves sent to zoos between 1978 and 1983. The calculation of vital rates and population growth rates between 1984 and 1988 does not include translocated adults.

Sex and Age Criteria and Breeding Status

We sexed animals by observing external genitalia, body size, size of neck folds (which are more pronounced in males), urination, and presence of calves. Also, the horns of adult males are wider at the base than those of adult females (chapter 4). Sex of calves younger than 1.5 years was difficult to determine, and we recorded it only if more than three observers had made the determination or if one observer had made it on three occasions. Laurie (1978) showed that researchers have incorrectly sexed a significant proportion of calves and that initial sightings are biased toward males. Thus we cautiously listed a large number of young calves as sex undetermined.

We classified rhinoceros in the Royal Chitwan National Park population as calves (4 years), subadults (4–6 years), or adults (6 years). The intensively studied Sauraha population provided more specific age categories: calves (0–1, 1–2, 2–3, 3–4 years); subadults (4–5, 5–6 years); young adults (6–12 years [i.e., breeding females and nonbreeding males]); intermediate-age adults (12–20 years); and older adults (20 years). We could age calves and subadults accurately because we knew the birth years of all calves and nearly all subadults. We could distinguish subadults from adults by body size and horn size. I based our age classification for subadults on the potential for animals to breed. Captive females and males show signs of reproductive activity by age six, so the subadult category consisted of animals aged four to six years. Young adults had molars with little wear, small lower incisors, short horns with little wear, and few scars or body marks and were small in size (chapter 4). Young adult males lacked pronounced secondary neck folds. We distinguished individuals in the twelve- to twenty-year category by moderate wear on the molars, horn size and wear, increased amount of facial wrinkles, size, scars, development of secondary neck folds in males, size of outer incisors, and for females, the birth of more than one calf. Individuals older than twenty had extensive wear on the molars and often displayed a combination of extensive facial wrinkles; major scars on the anal skin folds; torn or notched ears; broken, deeply grooved, or eroded horns; and, in males, extensive development of secondary neck and shoulder folds. We tested differences in age and sex classes by using chi-square analysis, with significance judged at the 0.05 level.

We identified males that were breeders by observing copulations ($N = 7$), tending of estrous females, the outcome of fights among dominant males, and behavioral and morphological features. Breeding males often squirt urine when closely approached, possess extensive secondary neck folds and large procumbent mandibular incisors, and are aggressive toward subordinate males (Laurie 1978).

Fecundity and Mortality Rates

We monitored the Sauraha population to estimate interbirth intervals using the period between births for eighty-seven registered breeding-age females. Zoo data showed that gestation in greater one-horned rhinoceros is about 15.7 months (Laurie, Lang, and Groves 1983). We assigned birth dates (± 1 calendar month) to calves born during the study in the Sauraha population. I calculated birthrates annually for each age class by dividing the number of live births within a female age class by the total number of females within that age class. Using data for the entire Sauraha population for both age-specific fertility and survival, I calculated rates each year for all individuals in each class and then averaged them over all years. Our fecundity schedule is based on age-specific births of female offspring.

Our study and official government records provided mortality data for the Sauraha population. Rhinoceros are considered the property of the king, and the park warden must account for all mortalities by an official inquest. Animals most frequently died close to riverbanks, and their carcasses attracted large flocks of vultures, aiding location of dead animals. Domesticated elephants frequently grazed in the same areas. It is unlikely that deaths, other than newborns, escaped detection. I calculated age-specific mortality by dividing the number of dead individuals within an age class by the total number within that class.

Population Growth Rate

I used two methods for estimating population growth rate: regression analysis on population estimates to give the observed rate of increase, and instantaneous rate of increase (Caughley's [1977:109] *r*, calculated from survival and fecundity tables. I made two assumptions: that the observed vital rates were fixed and that these rates had persisted long enough for the population to stabilize. A lack of annual data on vital rates between 1976 and 1984 hindered my attempts to determine the validity of the first assumption, especially the fixed mortality rates among older age classes. The longevity of rhinoceros and the difficulty of assigning animals to annual or biennial rather than broad age classes inhibited our effort to properly evaluate the assumption of a stable age distribution. However, several observations provide circumstantial evidence that the second assumption holds: (1) similarity in age structure of the Sauraha population in 1975 and 1988; (2) identical mortality rates for subadults and adults; and (3) similarity in birthrates between 1972 and 1975 and between 1984 and 1988. Laurie (1978) estimated birthrates with a different method, but his maximum estimate (8.9%) is close to our estimate of 7.6%. Continued census efforts in Sauraha will determine the validity of assuming a stable age distribution.

Genetic Analyses

We conducted genetic analyses in collaboration with Dr. Gary McCracken of the University of Tennessee. In 1986/1987, we obtained blood and dermal tissue samples from twenty-three individuals from Chitwan (about 6% of the current total population) that we had immobilized as part of a field study and translocation program. We separated plasma and red blood cells and froze all samples in liquid nitrogen. After the samples arrived in the laboratory, they were prepared for protein electrophoretic studies, as described in McCracken and Wilkinson (1988). Horizontal starch-gel electrophoresis were used, following the techniques of Selander et al. (1971) to examine twenty-nine presumptive protein-encoding loci: seventeen loci from dermal tissue (Aat, Es-I, 2, 3, 4, Fum, G3pdh, Gpi, Lap-1, 2, 'Ah-I, 2, Me, Pgm-1, 2, Pmi, and Sod); seven from red blood cells (Dia-1, 2, G6pdh, Hb, Mdh, Pep, 6Pgd); and five general proteins (Gp-1, 2, 3, 4, 5) from blood plasma. Aat, Mdh, Me, and Pmi were resolved using tris maleate buffCr5; Dia-1, 2, G6pdh, Gp-1-5, and Hb using lithium hydroxide buffers; Es-1-3, Fum, G3pdh, Lap-1, 2, and Ldh-1, 2 using tris-citrate buffers (pH 8.0); tris versene borate buffers to resolve Fs-4 and 6Pgd; and Poulik discontinuous buffers to resolve Gpi, Pep, Pgm-1, 2, and Sod (Selander et al. 1971). Protein stain recipes are from Selander et al. (1971) and Harris and Hopkinson (1978).

Habitat–Density Relationships

I used only 1988 data to compare densities within blocks with habitat variables because two variables—the percentage of cover of *Saccharum spontaneum* within the block and distance to riverbank—were subject to marked annual variation from siltation and flooding. I determined dominant plant associations within each block and the percentage of cover by *S. spontaneum* from aerial photographs and ground surveys. I used aerial photographs to measure distance from the edge of croplands to the center of the highest rhinoceros population density within a block (i.e., the part of the block containing the most frequent sightings of animals). Blocks ranged in size from 1.05 to 4.76 km^2, so we determined whether density was positively correlated with block size. We used Spearman rank correlation analysis to test for correlation between rhinoceros densities and habitat variables associated among blocks.

Home-Range Analysis and Movements

We used radiotelemetry to locate individuals and calculate home range and daily movements. For social organization and male dominance studies, we located animals twice a day. However, to avoid autocorrelation of location data in home-range analyses, I used only observations made at least forty-eight hours apart. This more strict criterion reduced the total locations to 926 during a three-year period for adult males ($N = 375$) and adult females ($N = 551$), and I calculated separately the seasonal home ranges and annual home range for adult males and adult females. For the calculation of seasonal home range, I used only individuals with more than 28 locations per season (monsoon season and cool-dry season) or 15 locations for the much shorter hot-dry season. Jnawali (1995) estimated that size of seasonal home range reached an asymptote after twenty-eight observations; my own observations showed that fifteen observations were suitable for calculating home range during the hot-dry season.

Once we located a radio-collared animal in the field, we marked its location on acetate overlaid on an aerial photograph of the study area. We later used these data in a Geographic Information Systems program (CAMRIS) (Ecological Consulting 1993). We used the harmonic isopleth method (Ecological Consulting 1993) rather than the more traditional minimum convex polygon method to calculate seasonal and annual home range. The latter method tends to overestimate home range in the case of rhinoceros in Chitwan because of the distribution of rhinoceros habitat. For example, the interdigitation in some agricultural enclaves causes the inclusion of areas that lie between natural habitats and that rhinoceros do not use. The harmonic isopleth better approximates the movements and locations of radio-collared individuals in these circumstances.

Diet Selection

Jnawali (1995) studied diet selection of the Chitwan population and the translocated Bardia population using fecal analysis. Jnawali and I worked together to build a reference slide collection of plant species, wild and cultivated, that rhinoceros eat. Jnawali extended his analysis of food habits to form the basis for his doctoral dissertation. Jnawali (1995) provides detailed descriptions of methodologies he used in the fecal analysis. For his study, he created a relative importance value for each species observed in the fecal sample, basing it on the mean percentage of species X in a fecal sample multiplied by the square root of the frequency of species X in the fecal sample. He used Ivlev's (1961)selectivity index to calculate selection of plants eaten.

Activity Patterns

Between 1984 and 1988, radio collars on eighteen adults (10 males, 8 females) allowed monitoring of their activity for twenty-four-hour cycles. We used trained elephants to follow focal animals. Before we conducted activity watches, we habituated all study animals to the presence of observers on elephant-back. As a result, rhinoceros behaved normally as they grazed and wallowed close to elephants. We used five-minute scan samples to record habitat selection, feeding, wallowing, and other behaviors during the twenty-four-hour activity watches. These totaled 14,842 observations (8,034 for females and 6,808 for males) spread over three seasons and three years.

Feeding behavior (see the section on diet selection) and wallowing were of particular interest. I did not use invasive techniques (i.e., implanting body temperature telemeters); instead, I recorded the number of minutes spent wallowing for twenty-four-hour periods and averaged this value for all twenty-four-hour watches conducted by month. In all, we conducted fifty-two twenty-four-hour watches, two to six each month. I recorded an individual as wallowing if it was immersed to its belly or deeper. I noted the length of time spent wallowing, time of day, interval between wallowing bouts, and location of wallow site (river or stream, grassland mud hole, riverine forest mud hole), and I divided the twenty-four-hour cycle into four periods: 0530–0930; 0931–1330; 1331–1730; and 1731–0530. I divided wallowing time per interval by interval length to account for unequal interval length.

I calculated monthly mean values for temperature and relative humidity by using weather data collected at the Rampur agricultural station on the border of Chitwan. I used monthly mean values because the weather data from Rampur were available for only part of this period. I converted the percentage of relative humidity values to vapor density (gm^3) based on average air temperatures (Campbell 1977). I used nonparametric correlation analysis (Wilkinson 1989) to determine relationships between environmental variables and wallowing time per month.

Dominance, Social Organization, and Reproduction in Males

To monitor movements and social organization among males, our fieldwork focused on Icharni Island, the area of highest density of breeding female rhinoceros in Chitwan. We radio-collared five of seven breeding males dominant during the four-year study interval and four other males present in the study area (Dinerstein, Shrestha, and Mishra 1990). Another male killed one dominant male (M039) before we could capture and radio-collar him. We frequently had to replace radio transmitters damaged as a result of fighting between males.

To examine the relationship between phenotypic variables and male behavior, I collected data on sedated animals or carcasses from the seven individuals that were dominant during the study period and from other breeding and nonbreeding males in adjacent areas. In particular, I compared incisor size, horn size, and body size of dominant males from Icharni with these attributes of males from elsewhere in the study area.

We twice recaptured four breeding males and recaptured one breeding male three times. Two of the seven individuals died from wounds suffered in fights with other males. We conducted twenty-four-hour activity watches on all five radio-collared males during each season and noted their locations on Icharni for fifteen to thirty days each month.

Early in my study, it was clear that the dominant male on Icharni Island was intolerant of other males in areas where females grazed extensively. I calculated how long a dominant male successfully excluded other males from becoming established on the island or remained superior to one or two other males present. Tenure length is the length (in days) between the first and last sightings of males on Icharni Island.

I used the presence of fresh cuts on the flanks, rump, and hind legs to establish changes in the status of males within the dominance hierarchy on Icharni. Such cuts were inflicted only when one male turned and ran from the other. This diurnal survey technique proved useful because fights commonly occur at night when it was difficult to stay close to and observe fighting males. Vocalizations of males uttered during nocturnal fights may carry for several kilometers. These served as a reminder to search the following morning to determine the outcome of the fight. The presence of scratches or wounds served as a bioassay for assigning dominance rank. I classified a male involved in a fight as dominant if he lacked new cuts on his flanks, rump, or hind legs and chased off other males. I considered males codominant if we observed fresh cuts but the male still tended females or remained on the periphery of Icharni. I considered males no longer dominant when they were forced off Icharni Island or if they took up residence elsewhere, especially if they moved to areas where reproductive-age females were less common.

To calculate breeding success for males on Icharni, I assumed that when a male was dominant, he was responsible for breeding all the females receptive during his tenure. This assumption seemed reasonable because I never observed a subordinate male chasing off a dominant male that was tending a female in estrus. Thus I was able to calculate a minimum estimate of total number of females bred. To estimate the maximum number of females bred, I assumed that males bred not only all females that conceived while that male was dominant but also all females in estrus while that male was codominant on Icharni.

Estimation of Breeding Schedules of Females

I determined the breeding schedules of females by subtracting gestation length (15.7 months) from birth dates. Observations of tending by a dominant male, estrus, or copulation also contributed to the estimation of breeding schedules. We radio-collared two-thirds of the adult females on Icharni to facilitate location and to monitor breeding activity at night or when vegetation obscured the animals. We also registered and photographed all females as part of a census begun in 1984 (chapter 5).

Because females that gave birth in the study area in 1984 had conceived in 1983, the analysis of reproductive schedules included 1983. We observed relatively few copulations, and they sometimes did not result in conception. Thus to estimate the number of conceptions that might have occurred in 1987 and 1988, we counted females as receptive if the current calf was between eighteen and twenty-four months old, the point at which most females, according to the estimated interbirth interval, are likely to breed again. We assigned paternity estimates only for females known to be resident on Icharni Island during the study. Jnawali and I, during my follow-up visit to Icharni in January 1990, provided birth data for females in the study area from June 1988 to February 1990.

Rhinoceros–Plant Interactions:
Is the Megafaunal Seed Dispersal Syndrome Valid?

We measured the daily consumption of *Trewia* by using twenty-four-hour activity watches of radio-collared rhinoceros. We conducted this and all other related activities and experiments between 1984 and 1986. When possible, we counted the number of fruits ingested during five-minute intervals by approaching habituated rhinoceros on elephant-back and making close observations through binoculars. When we could not accurately determine intake, we recorded only feeding. We estimated an index of monthly consumption by randomly selecting twenty 1-kg dung samples each month from grassland and forest latrines and counting the number of intact and cracked seeds present.

SEED PASSAGE EXPERIMENTS WITH A CAPTIVE RHINOCEROS. Every morning, we would collect the fruits in Chitwan that would be transported to the Kathmandu zoo the same day or the following day and fed to the captive rhinoceros. Rhinoceros regularly eat fallen fruit at least several weeks old, so it was safe to assume that a one-day delay in feeding would have no effect on the experiment. In the first feeding trial, the rhinoceros ingested 114 fruits during a ten-minute interval; in the second trial, the rhinoceros ate 300 fruits within fifty-four minutes.

After the rhinoceros ate the fruits, keepers collected the dung excreted in each bout and deposited it in numbered plastic bags, recording the date and time of defecation of each sample. We searched for and removed *Trewia* seeds the same day that the dung was excreted. The shiny black *Trewia* seeds stood out against the bright green excreta, so we missed few intact seeds passed by the rhinoceros. We also recorded fecal mass and number of intact and broken seeds passed in the dung.

Keepers continued to collect and we continued to examine fecal material for three days after the rhino excreted the last intact seed.

GERMINATION AND GROWTH EXPERIMENTS. To distinguish the effects of gut treatment from the effect of manuring on germination rate and seedling growth, I randomly assigned defecated and uningested seeds to be planted in dung and in seed-free soil. I conducted the experiment, a 2×2 factorial design, during the peak *Trewia* fruiting period in an open area. I housed randomly located replicates in 0.50-m² screened boxes that we protected from insect seed predators by frequently applying an effective chemical insecticide (Finit). We exposed intact seeds randomly selected from rhinoceros latrines and from entire ripe fruits to four treatments: sixty seeds defecated by rhinoceros and planted in 20 kg of dung; sixty seeds defecated by rhinoceros and planted in forest topsoil; sixty uningested seeds planted in dung; and sixty uningested seeds sown in forest topsoil. We replicated each treatment three times. Each day I noted the number of seeds germinated (i.e., cotyledon emerging from the seed coat). At the end of the growing season, fifty-one days later, I carefully removed the seedlings and recorded the wet mass of leaves, stems, and roots for each seedling. I dried plant parts at 70°C for forty-eight hours to obtain dry mass.

SEED SHADOWS AND DEFECATION. I determined daily movements and defecation patterns of rhinoceros by following radio-collared animals during twenty-four-hour periods. I marked hourly locations and defecation sites on acetate overlays of aerial photos of the study area. After each defecation, we placed the feces in a plastic bag and weighed it using a 10-kg Pesola spring scale.

FRUIT ROT AND THE FATE OF UNINGESTED FRUITS. I estimated the monthly availability of ripe and rotting fruits by counting all fruit on ten 100-m² forest plots and weighing ripe and rotten fruit separately. *Trewia* seeds emerge from the hard endocarp through ingestion or by being crushed when rhinoceros or elephant step on them. To determine the effects on seed germination of fruit crushing by large mammals, we observed and simulated (by having domesticated elephants walk over them) the extent to which rhinoceros and elephant crushed fruits. I randomly placed twenty 0.5-m² screened-in exclosures, ten in the forest and ten in short grassland. Using a randomized block design, I assigned 100 crushed fruits to one side of the divided box and 100 entire fruits to the other side. Each week, I recorded the number of seeds that germinated from each side of each replicate, and I recorded seedling height after fifty-one days.

LATRINE FLORAS AND RECRUITMENT OF *TREWIA* ON NATURAL LATRINES. I used the line intercept method (Canfield 1941) to measure plant cover at thirty-seven grassland latrines. In all, I sampled 306 m along intercepts. I stratified the sampling vertically to distinguish among seedling cover, sapling and shrub cover, and tree cover. Restrictions on sampling included using only active latrines larger than 1 m in diameter. Most latrines that we measured were elliptical and averaged 7 m along the longest axis. We conducted the sampling on August 28, when annual growth was nearing its peak. To evaluate recruitment on grassland and riverine forest latrines, we measured seedling number and height on randomly located square-meter plots on thirty grassland and thirty forest latrines.

RECRUITMENT OF TREWIA ON EXPERIMENTAL LATRINES. To evaluate light limitation as a mortality factor for *Trewia* seedlings, we created some experimental latrines: ten replicates in short and intermediate grasslands and ten under the riverine forest canopy. Each month we added 20 kg of fresh dung to each experimental latrine after we had deposited an initial base of 180 kg. These amounts seemed appropriate, based on initial observations of deposit patterns and size of natural latrines, although later measurements indicated that my estimates were conservative. Open-mesh fencing around experimental latrines excluded inter-mediate-size and the large herbivores and protected seedlings from fire. Records included seedling number and height for all seedlings in randomly located 4-m^2 plots on paired forest and grassland latrines at the end of two growing seasons.

To determine the effects of sun versus shade on seedling growth, I randomly lo-cated in various grassland types germination trials of 50 seedlings each in twenty "sun" latrines (equal to two defecations) and randomly located shade "shade" la-trines under the canopy of riverine forest. In each latrine, I planted 100 seeds in dung; when seedlings had produced three leaves, I weeded the plots to 50 individu-als, or a density of 1 seedling per 100 cm^2. I recorded mortality twice weekly until the end of the growing season (mid-November), at which time I measured the height of surviving seedlings.

RECRUITMENT UNDER THE CANOPY. *Trewia*-dominated riverine forest often occurs in distinct patches. I randomly located twenty 25- × 15-m plots among three of these patches and made records of diameter at breast height (dbh) of all trees, saplings, and woody climbers larger than 3 cm in diameter. I also tallied the diameters of all *Trewia* saplings larger than 2 m in diameter and the heights of all seedlings within the plots to allow classification of seedlings by age (1–3 years).

Rhinoceros–Plant Interactions: Testing for the Effects of Browsing on Riverine Forest Structure and Composition

To determine whether rhinoceros inhibit saplings of forage species from reaching the canopy, we looked for a control area where browse damage was not evident. Un-fortunately, we could not find such an area within Chitwan. We could not use forested areas adjacent to Chitwan because villagers use some of the same species for cattle fodder. Thus we used exclosures to assess the effect of rhinoceros on ver-tical growth of stems, using a haphazard sampling regime to establish study plots in areas where a 10- × 10-m exclosure or control plot would contain at least thirty-five *Litsea monopetala* stems, a preferred forage plant. A forest inventory revealed that such densities are typical in the riverine forest patches to the east of Icharni Is-land near Sauraha. We randomly assigned six plots as open to foraging rhinoceros and protected six plots from rhinoceros by erecting 2-m-high wooden fences. We located a control plot close to each exclosure. I initially marked and measured 180

Litsea saplings (15 per plot) in October 1985. At the time of exclosure, a nested one-way ANOVA (SAS 1989) showed that sapling height did not differ significantly between protected (grand mean ± 1 SD for six plots 2.49 $\pm$ 0.23 m, N = 90 saplings in six plots) versus unprotected saplings (2.46 $\pm$ 0.23 m, N = 90 saplings in six plots) (F 1,1 79 = 0.11 P = .75). At the end of three years, we used another nested one-way ANOVA (SAS 1989) to test for differences in sapling height.

To further evaluate the effect of browsing and trampling by rhinoceros, we inventoried all woody stems in ten 0.5-ha plots located in five patches of riverine forest (two plots per patch). We examined stems for characteristic browse marks and trampling by rhinoceros.

To determine whether rhinoceros browsing alters the amount of leaf tissue available below 2 m, we counted the number of branches that supported at least ten whole leaves below a height of 2 m (henceforth, leafy branches) on 337 saplings drawn from each of the twelve plots (N = 172 and 165 for protected and unprotected plots, respectively). I observed that rhinoceros could readily browse leaves below 2 m without trampling saplings. We marked the stems and followed them only during the first year. We used many saplings that we had marked to measure height also to measure distribution of leaves. We analyzed these data using a general linear model nested one-way ANOVA (SAS 1989) because of unequal replication of saplings within each plot.

Collection and Monitoring of Socioeconomic Data Related to the Hotel Industry

Where Ecotourism Dollars Go

Marnie Murray collaborated to develop the methods that we used for this aspect of the study. To assess the economic effect of ecotourism on the local economy, we studied the local hotel industry that was operating both inside and outside Chitwan. We developed three questionnaires, and six trained Nepalese surveyed hotel managers and employees in March 1995. A total of 144 interviews provided information on hotel ownership, package-tour rates (including food, lodging, and tourist activities for two days and three nights), visitation rates, employment levels, and employee salaries.

We then estimated the amount of revenue generated by the hotel industry for the 1994 tourist season to determine how much of the gross was recycled to the local community. We based this estimate on package-tour rates and visitor numbers recorded for each hotel in the hotel survey. To determine the contribution of different types of hotels (low-budget cottages, medium-price hotels, and expensive hotels) to local employment, we stratified the data collected from the hotel survey by package-tour rates and daily rates. We surveyed forty-nine hotels (three hotels were closed, and one hotel was under construction during the 1994 tourist season).

We also randomly interviewed 108 tourists staying at hotels in Sauraha and at the larger hotels inside the park. We designed the tourist questionnaire so that we could estimate visitors' spending on Chitwan excursions and identify trends in reservation bookings. We also sought information on tourists' willingness to pay so that we could calculate the park entry fee that would yield the greatest return.

Earning Capacity of Nature Guides

The demand for and employment of local nature guides in Chitwan is one of the most beneficial contributions of ecotourism to the local economy. Although hotels employ the majority of guides, many guides freelance. Therefore, to capture complete information on the effects of ecotourism on the livelihoods of local residents, we had to conduct a separate study of nature guides. We surveyed the nature guides in August 1995 to determine the average annual salary they were commanding in the privately owned ecotourism industry. We visited all hotels in the area and interviewed 140 nature guides working in the local ecotourism industry (65% of the estimated number of trained guides in the Chitwan vicinity).

Effects of Privately Owned Ecotourism on Household Income

To determine the distribution and magnitude of economic benefits from ecotourism to local villagers, we collected information on both direct and indirect household income generated from work or activities related to ecotourism. We randomly surveyed the residents of 996 houses in fifty-seven wards of seven village committees located along the periphery of Chitwan, or approximately 9% of the households in those seven village committees. The seven village committees in our study area were Bachhauli, Bhandara, Kathar, Khairahani, Kumrose, Patihani, and Piple. We chose these village committees because of their proximity to the main entrance of Chitwan in Sauraha, the center of ecotourism activities. All seven village committees were within 2 km of the park boundary and within 15 km of Sauraha. We defined direct economic benefits as personal income generated from employment in the industry (e.g., nature guiding, elephant driving, or cooking for hotels) and indirect economic benefits as total revenue (price multiplied by quantity sold) from the sale of products or the provision of services related to ecotourism (e.g., souvenir sales, cultural dance performances, or independent guiding separate from hotels). We used the data to cross-check salary estimates reported by hotel owners in the hotel survey.

Methods for Estimating Trajectories of Sensitive Species, Processes, and Landscape-Scale Conservation

We set up a monitoring program to estimate the number of target species that recolonized the regenerating (previously degraded) forests. We relied on photographic identification of rhinoceros (Dinerstein and Price 1991), pug mark and scat analyses for tigers and tiger prey, and direct observations of all three species. We monitored recruitment by direct observation (rhinoceros) and pug marks (tigers). We conducted censuses twice a month from elephant-back in the two main regeneration areas; each census used the same observers and as many as four elephants per census. We also included a control plot in the census: Icharni Island, an adjacent block of habitat inside Chitwan that supports one of the highest recorded densities of rhinos anywhere (Dinerstein and Price 1991). We also monitored the relative abundance of tiger prey in regeneration areas.

We are continuing to collect data on the poaching of rhinoceros and tigers; we present here only data from January 1990 to March 1997. For rhinoceros, the data come mostly from field notes or conversations with the senior and assistant park wardens: Ram Pritt Yadav and Tika Ram Adikhari. Both officials lived in Chitwan during much of the monitoring period, and Yadav has been a warden since 1976. We also took global positioning satellite (GPS) readings of each poaching incident and used a geographic information system program (Ecological Consulting 1993) to display poaching incidents on a yearly basis in and around Chitwan. To analyze the distribution of poaching events and identify poaching hot spots, we used a harmonic mean isopleth analysis program (Ecological Consulting 1993) for the forty-five poaching incidents that occurred between 1990 and 1997.

To assess how project interventions led to the regeneration of critical habitats preferred by endangered species, we used GIS to map the project area and calculated the size of restored blocks of habitat. We also monitored the extent to which user group committees followed restoration recommendations made by the staff of the King Mahendra Trust. These recommendations included excavating wallows for rhinoceros, fencing tall grassland habitat adjacent to riverine forests (Kumrose only), stopping the collection of firewood from inside the regeneration areas, managing tourism in the wildlife-viewing areas, maintaining shortgrass clearings for tiger prey (Bagmara), and leaving standing dead trees (snags) for cavity-nesting birds.

Direct measurement of the maintenance or restoration of ecological processes is often difficult. For example, intensive radiotelemetry studies are the best way to monitor dispersal of tigers, an important ecological process. We chose a surrogate measure, the restoration of habitat integrity in wildlife corridors, a prerequisite for effective dispersal. We also assessed predation by tigers on native herbivores in regeneration areas by counting kills. We used the presence of tigers, their prey, and breeding rhinoceros, taken together, to gauge the maintenance of early successional habitats containing the full suite of native large mammal species.

MEASUREMENTS AND OTHER PHYSICAL FEATURES OF GREATER ONE-HORNED RHINOCEROS CAPTURED IN ROYAL CHITWAN NATIONAL PARK

TABLE B.1. Univariate Statistics of Physical Characteristics (cm) of *Rhinoceros unicornis* of Various Ages, Royal Chitwan National Park, 1985–1988

Character	Calves <1 year old			Subadults			Young adults			Older adult males			Older adult females		
	N	$\bar{X}$	SD	N	$\bar{X}$	SD	N	$\bar{X}$	SD	N	$\bar{X}$	SD	N	$\bar{X}$	SD
Total length of body	7	185.3	37.4	6	351.2	14.3	10	386.8	12.0	15	411.7	20.6	9	399.2	24.7
Length of tail	7	34.1	3.2	6	56.5	7.0	10	63.4	3.7	15	65.7	5.1	9	63.9	10.0
Length of head and body	7	151.1	35.5	6	294.7	15.5	10	323.3	9.7	15	346.0	18.4	9	335.3	25.7
Length of hind foot to base of middle hoof	7	29.3	5.0	6	45.5	6.8	7	47.7	1.6	14	49.3	2.7	7	47.7	4.1
Length of hind foot to tip of middle hoof	7	32.9	3.4	6	51.2	3.4	7	51.4	2.4	14	55.6	4.0	7	53.1	4.1
Length of ear	7	16.0	1.9	6	24.2	1.3	10	25.0	1.0	15	24.7	0.9	9	25.9	1.7
Maximum skull circumference	7	89.4	15.0	6	138.5	8.1	9	151.9	9.9	10	170.7	10.5	9	157.0	8.2
Chest circumference	3	116.7	12.5	3	252.0	34.8	3	275.0	29.2	4	314.5	28.1	3	298.7	41.5
Neck circumference behind head	7	73.6	13.2	6	112.2	12.8	9	134.4	14.4	13	159.2	12.6	9	135.7	5.8
Neck circumference in front of shoulder	7	94.1	21.0	6	144.5	18.9	9	159.3	12.2	13	201.3	16.7	9	173.0	8.5
Shoulder height	3	72.7	2.6	4	155.8	18.8	3	150.3	21.9	4	172.3	14.2	3	149.3	14.7
Length of head	5	38.6	4.5	2	49.0	17.0	3	64.01	1.4	8	68.8	2.9	5	69.2	8.1
Cranial breadth	6	25.3	2.5	3	36.3	2.1	6	38.3	1.3	11	42.0	1.3	7	40.6	1.5
Width behind head	6	24.2	4.1	3	32.7	1.7	6	36.5	3.7	11	47.5	5.9	7	41.4	1.8
Width in front of shoulder	6	26.3	6.2	3	43.3	5.7	6	45.5	5.4	11	55.6	8.1	7	49.6	4.0
Shoulder width	5	34.8	6.1	3	44.7	10.0	6	38.3	1.2	11	67.2	11.6	7	59.7	4.0
Width across anterior cross-skin fold	5	43.2	7.3	3	73.0	7.1	5	78.2	5.8	8	86.1	5.2	7	86.6	3.7
Width across posterior cross-skin fold	5	41.0	7.8	3	74.3	3.9	5	80.8	5.0	11	87.4	4.1	7	90.3	6.4
Length of horn	—	—	—	6	12.2	2.0	9	15.94	3.5	15	25.3	4.7	9	23.8	4.6
Horn circumference at base	—	—	—	2	41.5	1.5	7	43.3	5.2	11	59.0	12.0	9	46.2	3.0
Length of lower left outer incisor	5	0.8	0.3	6	1.2	0.6	9	3.0	1.7	16	5.4	2.7	9	4.2	0.7
Length of lower right outer incisor	5	0.8	0.3	6	1.3	0.5	9	3.0	1.8	14	5.6	2.4	9	4.4	0.8

TABLE B.2. PROPORTION OF ADULT *RHINOCEROS UNICORNIS* EXHIBITING VARIOUS CHARACTERISTICS, ROYAL CHITWAN NATIONAL PARK, 1985–1988

Character and sample size	Young females (N=39)	Young males (N=32)	Intermediate-aged females (N=33)	Intermediate-aged males (N=23)	Old females (N=27)	Old males (N=36)
Knobs on second cross-skin fold	0.18	0.06	0.30	0.22	0.44	0.28
Knobs on anal skin fold	0.08	0.06	0.27	0.22	0.26	0.28
Knobs elsewhere	0.05	—	0.03	—	—	0.08
Scars on first cross-skin fold	0.03	—	0.03	0.04	0.11	—
Scars on second cross-skin fold	0.08	0.09	0.12	0.04	0.15	0.11
Horn broken off	—	—	0.06	—	0.59	0.31
Horn heavily eroded at base	—	—	0.15	0.04	0.56	0.19
Horn with longitudinal groove in front	0.08	0.13	0.45	0.43	0.04	0.39
Pigmentation	—	—	—	—	—	—
First extra cross-skin fold	0.03	—	—	—	—	0.03
Second extra cross-skin fold	0.03	0.03	0.06	—	0.07	—
Extra neck-skin fold (females only)	0.13	—	0.24	—	0.19	—
One ear cut	0.03	0.13	0.15	0.22	0.15	0.53
Two ears cut	—	0.03	—	—	—	0.11
Tail tip missing or bent	—	—	0.12	—	0.11	0.06

DEMOGRAPHIC AND GENETIC DATA

TABLE C.1.   ABRIDGED LIFE TABLE FOR FEMALE GREATER
ONE-HORNED RHINOCEROS IN THE SAURAHA SUBPOPULATION, ROYAL
CHITWAN NATIONAL PARK, BASED ON MEAN MORTALITY RATES

Age interval (yr)	l_x[a]	p_x[b]	q_x[c]	m_x[d]
0–1	1.000	0.889	0.111	0.0000
1–4	0.889	1.000	0.000	0.0000
4–6	0.889	0.950	0.050	0.0000
6–12	0.802	0.988	0.012	0.0845
12–20	0.746	0.992	0.008	0.0915
20–35	0.700	0.989	0.011	0.0695
35[e]	0.593	0.000	1.000	0.0000

[a] l_x refers only to survivorship up to beginning of an interval.
[b] p_x is the average survivorship within an interval.
[c] m_x is the mortality rate.
[d] m_x is the number of female offspring per female per time unit.
[e] is estimated mean maximum age based on captive animals; females probably do
not continue to breed in the wild beyond thirty-five years.

SOURCE: After Caughley (1977).

TABLE C.2. SEX AND AGE STRUCTURE OF GREATER ONE-HORNED RHINOCEROS POPULATIONS IN THREE AREAS IN ROYAL CHITWAN NATIONAL PARK, APRIL 1988

Age category (yr)	Sauraha						West				Bandarjhola and Narayani River			
		Relocated		Relocated	Sex				Sex				Sex	
	M	M	F	F	unknown	Total	M	F	unknown	Total	M	F	unknown	Total
Calves														
(0–1)	4	—	2	1	8	15	—	—	—	—	—	—	—	—
(1–2)	3	2	2	3	4	14	—	—	—	—	—	—	—	—
(2–3)	4	1	3	1	9	18	—	—	—	—	—	—	—	—
(3–<4)	5	—	3	1	2	11	1	1	6	8	2	1	2	5
(0–<4) combined	19	—	16	—	23	58	1	1	6	8	2	1	2	5
Subadult														
(4–5)	9	—	3	1	3	16	—	—	—	—	—	—	—	—
(5–<6)	3	1	7	—	6	17	1	1	2	4	—	—	—	—
(4–<6) combined	13	—	11	—	9	33	1	1	2	4	3	4	3	10
Adult														
(6–12)	23	3	37	5	0	60	9	6	7	22	3	1	1	5
(12–20)	16	—	28	2	0	46	2	9	2	13	2	5	—	7
(>20)	19	2	22	1	0	44	3	—	—	3	7	—	—	7
(6–>20) combined	63	—	95	—	0	158	14	15	9	38	12	6	1	19

TABLE C.3. AGE-SPECIFIC BIRTHRATES FOR ADULT FEMALES, SAURAHA SUBPOPULATION OF GREATER ONE-HORNED RHINOCEROS, ROYAL CHITWAN NATIONAL PARK, 1984–1988

Age category (yr)	Census year				Mean	SE
	1984–1985	1985–1986	1986–1987	1987–1988		
6–12	0.179	0.310	0.138	0.048	0.169	0.048
12–20	0.167	0.233	0.167	0.167	0.183	0.015
>20	0.208	0.130	0.130	0.087	0.139	0.022
No. calves born/yr	13.000	18.000	16.000	18.000	16.300	1.025

TABLE C.4. ALLELE FREQUENCIES AT THE POLYMORPHIC LOCI EXAMINED IN GREATER ONE-HORNED RHINOCEROS

Polymorphic loci			Number of
Locus	Allele	Frequency	individuals examined
Es-3	a	0.05	21
	b	0.59	
	c	0.36	
Es-4	a	0.84	22
	b	0.16	
G6pdh	a	0.50	20
	b	0.50	
Gp-3	a	0.11	19
	b	0.89	
Gpi	a	0.80	23
	b	0.20	
Hb	a	0.93	23
	h	0.07	
Ldh-1	a	0.48	23
	b	0.52	
	c	0.02	
6Pgd	a	0.04	23
	b	0.94	
Pgm-2	a	0.17	23
	b	0.83	

NOTE: Dia-1 also was variable with a single rare allele at 0.02.

SEASONAL HOME RANGE
AND DAILY MOVEMENTS

TABLE D.1. SEASONAL HOME-RANGE SIZES OF GREATER ONE-HORNED RHINOCEROS IN ROYAL CHITWAN AND BARDIA NATIONAL PARKS

Season/sex		Number of locations	Area (km²) Core area: 50% of locations	Home range: 95% of locations	Minimum convex hull	Percentage of locations in core area in *Sa. sp.*	Percentage of locations in home range in *Sa. sp.*	Percentage of locations in home range in RF
Monsoon season								
Adult females								
F010	Gajuri	61	0.45	1.68	1.66	64.3	55.7	41.0
F002	Khagchiruwa	57	0.52	2.85	2.42	64.8	61.4	36.8
F004	Tindharke	44	0.35	3.53	3.91	65.5	75.0	20.5
F014	Laxmi	44	0.22	2.09	2.14	49.1	52.3	43.2
F003	Abire Pothi	65	0.28	1.39	1.86	88.6	63.1	36.9
Mean ± SD			0.4 ± 0.1	2.3 ± 0.9	2.4 ± 0.9	66.5 ± 14.1	61.5 ± 8.7	35.7 ± 8.9
Breeding adult males								
M038	Conan	43	0.28	1.93	2.08	48.7	62.8	30.2
M008	Karne	47	0.49	4.23	3.13	46.6	40.4	48.9
M009	Yadav	48	1.74	6.21	5.97	29.4	29.2	20.8
Mean ± SD			0.8 ± 0.8	4.1 ± 2.1	3.7 ± 2.0	41.6 ± 10.6	44.1 ± 17.1	33.3 ± 14.3
Cool season								
Adult females								
F010	Gajuri	42	0.37	2.26	2.19	62.1	64.3	31.0
F002	Khagchiruwa	24	0.85	3.51	2.34	60.0	66.7	33.3
F004	Tindharke	32	0.91	5.05	3.64	57.0	56.3	31.3
F014	Laxmi	11	1.17	4.28	1.75	39.3	54.5	45.5
F003	Abire Pothi	19	0.25	3.24	2.09	44.1	52.6	47.4
Mean ± SD			0.7 ± 0.4	3.7 ± 1.1	2.4 ± 0.7	52.5 ± 10.2	58.9 ± 6.2	37.7 ± 8.1

Breeding adult males								
M038	Conan	23	0.22	2.02	1.61	24.6	30.4	56.5
M008	Karne	32	0.44	1.87	2.01	47.8	43.8	53.1
M009	Yadav	27	2.26	5.98	4.98	23.2	14.8	3.7
Mean ± SD			1.0 ± 1.1	3.3 ± 2.3	2.9 ± 1.8	31.9 ± 13.8	29.7 ± 14.5	37.8 ± 29.6
Hot season								
Adult females								
F010	Gajuri	24	0.27	2.53	1.86	76.4	75.0	20.8
F002	Khagchiruwa	23	0.38	3.01	1.40	89.8	78.3	21.7
F004	Tindharke	15	1.24	7.38	5.14	53.4	60.0	13.3
F014	Laxmi	13	2.02	8.44	3.59	37.4	46.2	38.5
F003	Abire Pothi	17	0.52	1.70	1.29	73.8	58.8	35.3
Mean ± SD			0.9 ± 0.7	4.6 ± 3.1	2.7 ± 1.7	66.2 ± 20.7	63.6 ± 13.1	25.9 ± 10.6
Breeding adult males								
M038	Conan	20	0.59	3.78	3.25	32.6	45.0	40.0
M008	Karne	22	0.49	2.54	2.16	95.6	72.7	27.3
Mean ± SD			0.5 ± 0.1	3.2 ± 0.9	2.7 ± 0.8	64.1 ± 44.5	58.9 ± 19.6	33.6 ± 9.0

NOTE: Core areas represent 50% of all locations; home range is 95% of all locations. Both calculations derived using harmonic surface isopleth method (Ecological Consulting, Inc. 1993); convex hull method included for comparison. *Sa. sp*, *Saccharum spontaneum* grasslands; RF, riverine forest.

TABLE D.2. Daily Movements of Habituated Individual Rhinoceros During the Twenty-four-Hour Cycle in Royal Chitwan National Park

Hot-dry season	Date	Cool-dry season	Date	Distance traveled (km)	Monsoon season	Date	Distance traveled (km)	Mean distance traveled daily regardless of season
Adult females								
F010	02/13/88	F010	11/22/88	3.8	F003	9/11/86	5.1	—
F010	02/22/88	F010	12/23/85	2.7	F003	9/16/86	4.9	—
F010	03/03/88	F010	12/29/85	3.6	F003	9/19/86	4.8	—
F010	02/06/88	F004	12/08/85	4.6	F003	10/05/86	5.6	—
F010	04/10/88	F003	11/14/86	1.7	F003	10/17/87	4.9	—
F005	04/11/88	F003	11/25/86	5.1	F004	10/31/87	4.3	—
F005	04/18/88							
F003	04/10/87							
F003	05/31/87							
Mean				3.58			4.93	4.64
SD				1.24			0.42	2.02
Adult breeding males								
		M038	1/22/86	3.2	M038	7/10/86	5.8	—
		M009	12/15/85	12.5	M009	8/7/85	6.0	—
		M009	1/8/86	5.8	M009	8/11/85	5.7	—
					M009	9/28/85	5.5	—
Mean				7.17			5.75	6.36
SD				4.80			0.21	2.89

A PROFILE OF RHINOCEROS BEHAVIOR

This brief annotated ethogram of rhinoceros behaviors draws on Laurie's (1978) study and my observations. These are particularly relevant to the themes of this book. Readers interested in comparing the ethology of greater one-horned rhinoceros with that of other rhinoceros species should consult Laurie (1978), as he covers this topic in detail.

Associations

Social Groupings

- Solitary males. Adult males are almost always solitary.
- Adult females with calves. Adult females with calves do not form permanent associations with other individuals, regardless of sex and age. An adult female will allow an older calf to accompany her and her newborn calf; on some occasions, a female would drive the older calf away. Adult females with calves often associate with other females at prime wallowing areas or in concentrations of new shoots of *Saccharum spontaneum*. However, these associations are temporary (see the section on aggregations).
- Adult females without calves. Adult females without calves are largely solitary.
- *Subadult males and subadult females*. The most common and seemingly stable grouping is among subadults, particularly subadult males. Subadult males form groups of two to three individuals on the periphery of the core home ranges of dominant

males, probably invoking the strategy of safety in numbers to protect themselves or to more easily detect these highly aggressive individuals. It would be interesting to apply genetic analyses to social groupings to learn the degree of relatedness of the subadults that cluster together. Subadult females are only slightly less social than subadult males. The frequency with which subadult females associated in my study area facilitated translocation programs because subadult females often served as founders in new populations.

Aggregations

Aggregations are, by definition, short-term groupings of individuals. The most common aggregations occur during the monsoon season in forest wallows and in *Saccharum spontaneum* grasslands during March and early April. In a forest wallow that measured about 30 m × 6 m, I observed as many as nine individuals, with other animals nearby. Aggregations in wallows typically included a dominant male, cow–calf pairs, and single adult or subadult females. I never observed dominant males sharing wallows with a subadult male. Once I counted thirty individuals in an area of about 0.25 km^2 (North Pipariya grassland in late March 1986) where all individuals were very near one another. This stands as a record to date. Most associations in prime grazing areas rarely exceed six to ten individuals, and the large majority of these were cow–calf pairs.

Communication

Rhinoceros communicate using at least ten vocalizations and by urine, feces, and scent.

Vocalizations

- *Snort.* An explosive sound often made on first encounter with another rhinoceros, either approaching or being approached. Laurie (1978) observed the number of snorts to range from one to twenty, averaging about two per episode.
- *Honk.* A loud vocalization that travels over long distances, often uttered during head-to-head confrontations or typically during the ensuing flight and chase of one animal by the other. In most cases, the loser of these encounters does the most honking.
- *Bleat.* A sound associated with submission during head-to-head encounters or during flight, often during courtship chases.
- *Roar.* A vocalization confined to intense encounters, both between males and females and between females. I also observed roaring when domesticated elephants deliberately or accidentally separated a very aggressive female from a calf younger than one year.

- Squeak-pant. One of the most charming and incongruous vocalizations uttered, typically by males during courtship chases.
- Moo-grunt. Essentially a low-intensity contact call between mothers and calves that, unlike the other calls, can be heard for only a short distance.

Other, much less common, vocalizations include shrieks, groans, rumbles, and humphs.

Olfactory Communication

URINATION. Few displays are as spectacular as the squirt-urination that dominant males practice. Although both sexes are capable of squirt-urination, dominant males regularly eject urine as far as 3 to 4 m behind them, whereas females did so only during apparent periods of estrus. Dominant bulls perform this behavior often when approached by an observer on elephant-back. Squirt-urination is also associated with other behaviors in males, such as foot dragging and horn and head rubbing on vegetation. Laurie (1978) identified seven contexts associated with urination, ranging from disturbance by an observer (the most common) to entering water.

DEFECATION. Observations of defecation were an important part of this study because of the ecological effects of this behavior on the dispersal and recruitment of the seeds of fruits ingested by rhinoceros. Both Laurie (1978) and I observed that the majority of defecations occur within 10 m of existing latrines. A mother's defecation often stimulates her calf to do so. For detailed information on the ecology of dung piles, see Dinerstein (1992). A summary appears in chapter 8.

SCENTS. Rhinoceros reacted to the scent of other conspecifics by sniffing the ground or vegetation where other males had sprayed urine. They typically engaged in flehmen, also known as lip curl, a behavior shown by both males and females; rhinos also use this to assess reproductive status through smelling urine. Rhinoceros possess pedal scent glands, which they use to mark their presence at or in the vicinity of latrines.

During twenty-four-hour observations of individuals, habituated dominant males sometimes walked with their heads to the ground to follow what appeared to be female scents.

Interactions

Peaceful Interactions

Rhinoceros engage in a number of peaceful interactions, the most common of which is rubbing flank against flank or placing the head on the other's flank or head, behaviors typical of mothers and calves. Other behaviors in this category include greeting with head waving or bobbing, mounting flanks or rump, licking, nose-to-nose nuzzling, horn-to-horn sparring, playfully running around, playing with twigs in mouth, and other forms of unaggressive physical contact.

Agonistic Interactions

Laurie (1978) identified eleven components of agonistic behavior: the sudden turn, prolonged stare, lip curl, advancing steps, charge, horn-to-horn stare, tusk display, horn clash and lunge, tail curling, immediate flight, and prolonged chase. Agonistic behavior most commonly occurs between adult males and females with calves. Males frequently attack subadult males, and fights between dominant males are the most common source of mortality (chapter 5). Younger males do not exhibit agonistic behavior, whereas such behavior is common among dominant males.

Interactions during courtship are often agonistic — males chase females for long distances and sometimes engage in face-to-face combat. Both sexes vocalize during chases.

REPRODUCTIVE HISTORIES OF
ADULT FEMALE RHINOCEROS

TABLE F.1. Reproductive Histories of Resident and Transient Adult Female Greater One-Horned Rhinoceros on Icharni Island, Royal Chitwan National Park, January 1, 1983–June 1, 1988

Number and name	First registered calving date	Date of conception	Second registered birth date	Date of conception	Estimated date of next conception	Known no. of conceptions during study	Estimated maximum no. of conceptions during study
Resident females on Icharni							
F001 RC Khag Chiruwa	8/84	4/83	1/88	8/86	3/88[a]	1	2
F003 RC Abire Pothi	4/86	12/84	8/89[b]	4/88		2	2
F004 RC TinDharke	9/85	5/84	7/87	3/86	3/88	2	3
F005 RC Gumauri Pothi	12/85	8/84			5/87[b]	1	2
F007 Thuli Pothi					1/88[a]		
F010 RC Gajuri Pothi	5/84	1/83	10/86	6/85	6/87[a]	1	2
F018 Saanokhag Pothi	8/84	9/82	8/89[b]	4/88		1	1
F041 Jawaani NeckHole				2/88?		1	
Translocated resident females							
F011 Kunde Pothi	3/84	11/82					
F017 KalilobacchakoMau Pothi	11/84	7/83					
F040 BhayaaDholo Pothi							
F013 Tikkokanchi				1/86[a]	1		

Transient females resident on Icharni during conception period

ID	Name						
F014 RC	Laxmi Pothi	5/86	1/85	1/88	9/86	2	2
F073	Mystery Pothi	1/86	9/84[a]	1/88	9/86	2	2
F031	Shanti Pothi	7/84	3/83	2/87	10/85	1	1
F009 RC	Yalli Pothi	12/86	8/85	8/87?		1	2
	Total					14	21

Transient females not resident on Icharni during conception period and rarely sighted on Icharni

ID	Name	Unknown					
F002	Tutti Pothi	Unknown					
F034	Duifold	6/85	2/84				1
F026 RC	HaaneAuneMau	1/85	7/83	4/88	8/89[b]	11	1
F043	Ghoto Dholo	8/85	4/84	4/88	8/87?	1	2

NOTE: RC, radio-collared during part of the period 1984–1988.

[a] Male observed tending female or other evidence of copulation.

?, Female likely was receptive, given the age of calf.

[b] Calving date confirmed by research staff of Smithsonian/Nepal Terai Ecology Project.

REFERENCES

Alexandré, D. Y. 1978. Le rôle disseminateur des éléphants en forêt de Tai, Cote D'Ivoire. *Terre et la Vie* 32:47–71.

Allendorf, F. W. 1986. Genetic drift and the loss of alleles versus heterozygosity. *Zoo Biology* 5:181–90.

Allendorf, F. W., and R. F. Leary. 1986. Heterozygosity and fitness in natural populations of animals. In M. E. Soulé, ed., *Conservation Biology: The Science of Scarcity and Diversion,* pp. 57–76. Sunderland, Mass.: Sinauer.

Alverson, W. S., D. M. Waller, and S. L. Solheim. 1988. Forests too deer: Edge effects in northern Wisconsin. *Conservation Biology* 2:348–58.

Amato, G. A., D. Wharton, Z. Z. Zainuddin, and J. R. Powell. 1995. Assessment of conservation units for all Sumatran rhinoceros. *Zoo Biology* 14:395–402.

Andau, M. P. 1987. Conservation of the Sumatran rhinoceros in Sabah, Malaysia. *Proceedings of the Fourth IUCN/SSC Asian Rhino Specialist Group Meeting, Rimba, Indonesia* 21: 39–45.

Andau, M. P. 1995. Letter to editor. *Conservation Biology* 9:980–81.

Ayala, F. J., et al. 1972. Enzyme variability in the *Drosophila willistoni* group IV: Genetic variation in populations of *Drosophila willistoni*. *Genetics* 70:113–39.

Beier, P., and R. F. Noss. 1998. Do habitat areas and habitat corridors provide connectivity? *Conservation Biology* 12:1241–52.

Bell, R. H. V. 1971. A grazing ecosystem in the Serengeti. *Scientific American* 225:86–93.

Berger, J. 1986. *Wild Horses of the Great Basin: Social Competition and Population Size*. Chicago: University of Chicago Press.

Berger, J. 1990. Persistence of different-sized populations: An empirical assessment of rapid extinctions of bighorn sheep. *Conservation Biology* 4:91–98.

Berger, J. 1994. Science, conservation, and black rhinos. *Journal of Mammalogy* 75: 298–308.

Berger, J. 1999. Intervention and persistence in small populations of bighorn sheep. *Conservation Biology* 13:432–36.

Berger, J., and C. Cunningham. 1996. Is rhino dehorning scientifically prudent? *Pachyderm* 21: 60–68.

Berwick, S. H. 1974. The community of wild ruminants in the Gir Forest ecosystem, India. Ph.D. diss., Yale University.

Biodiversity Conservation Network. 1995. *Evaluating an Enterprise-Oriented Approach to Community-Based Conservation in the Asia/Pacific Region*. Washington, D.C.: Biodiversity Support Program.

Bonnell, M. L., and R. K. Selander. 1974. Elephant seals: Genetic variation and near extinction. *Science* 184:908–9.

Bookbinder, M., et al. 1998. Ecotourism's support of biodiversity conservation. *Conservation Biology* 12:1399–1404.

Borner, M. 1979. A field study of the Sumatran rhinoceros, *Dicerorhinus sumatrensis* (Fischer 1814). Ph.D. diss., University of Basle, Switzerland.

Brett, R. 1998. Mortality factors and breeding performance of translocated black rhinos in Kenya, 1984–1995. *Pachyderm* 26:69-82.

Brooks, M. 1994. Chairman's report: African Rhino Specialist Group. *Pachyderm* 18:16–18.

Bruner, A. G., R. E. Gullison, R. E. Rice, and G. A. B. da Fonseca. 2001. Effectiveness of parks in protecting tropical biodiversity. *Science* 291:125–28.

Burkhill, H. R. 1910. Notes from a journey to Nepal. *Records of the Botanical Survey of India* 4:59–140.

Buss, I. O., and J. A. Estes. 1971. The functional significance of movements and positions of the pinnae of the African elephant. *Journal of Mammalogy* 52:21–27.

Calder, W. A., III. 1984. *Size, Function, and Life History*. Cambridge, Mass.: Harvard University Press.

Campbell, G. S. 1977. *An Introduction to Environmental Physics*. New York: Springer-Verlag.

Canfield, R. H. 1941. Application of the line interception method in sampling range vegetation. *Journal of Forestry* 39:388–94.

Caughley, G. 1969. *Wildlife and Recreation in the Trisuli Watershed and Other Areas in Nepal*. Trisuli Watershed Development Project, Report no. 6. Kathmandu: Nepal Food and Agriculture Organization and United Nations Development Program.

Caughley, G. 1977. *Analysis of Vertebrate Populations*. New York: Wiley.

Cave, A. J. E., and D. B. Allbrook. 1958. Epidermal structures in a rhinoceros (*Ceratotherium simum*). *Nature* 182:196–97.

Chakraborty, R., P. A. Fuerst, and M. Nei. 1980. Statistical studies on protein polymorphism in natural populations III: Distribution of allele frequencies and the number of alleles per locus. *Genetics* 94:1039–63.

Chapman, J. A., and J. E. C. Flux, eds. 1990. *Rabbits, Hares, and Pikas: Status Survey and Conservation Action Plan.* Gland, Switzerland: International Union for Conservation of Nature.

Clutton-Brock, T. H., ed. 1988. *Reproductive Success.* Chicago: University of Chicago Press.

Department of National Parks and Wildlife Conservation. 1996. Survey of tiger populations in Terai parks and reserves. Babar Mahal, Nepal.

Dhungel, S. K., and B. W. O'Gara. 1991. *Ecology of the Hog Deer in Royal Chitwan National Park, Nepal.* Wildlife Monographs, no. 119. Bethesda, Md.: Wildlife Society.

Dinerstein, E. 1975. Vegetation of the Rapti River floodplain, Royal Chitwan National Park, Nepal. *Nepal Journal of Forestry* 4:16–23.

Dinerstein, E. 1979. An ecological survey of the Royal Bardia Wildlife Reserve, Nepal. Part I: Vegetation, modifying factors, and successional relationships. *Biological Conservation* 15:127–50.

Dinerstein, E. 1980. An ecological survey of the Royal Karnali–Bardia Wildlife Reserve, Nepal. Part III: Ungulate populations. *Biological Conservation* 18:5–38.

Dinerstein, E. 1985. Just how many rhinos are there in Chitwan? *Smithsonian Nepal Terai Ecology Project Newsletter* 2:5.

Dinerstein, E. 1987. Deer, plant phenology, and succession in the lowland forests of Nepal. In C. Wemmer, ed., *Biology and Conservation of the Cervidae*, pp. 272–88. Washington, D.C.: Smithsonian Institution Press.

Dinerstein, E. 1989. The foliage-as-fruit hypothesis and the feeding ecology of south Asian ungulates. *Biotropica* 21:214–18.

Dinerstein, E. 1991a. Sexual dimorphism in greater one-horned rhinoceros (*Rhinoceros unicornis*). *Journal of Mammalogy* 72:450–57.

Dinerstein, E. 1991b. Seed dispersal by greater one-horned rhinoceros (*Rhinoceros unicornis*) and the flora of *Rhinoceros* latrines. *Mammalia* 55:355–62.

Dinerstein, E. 1992. Effects of *Rhinoceros unicornis* on riverine forest structure in lowland Nepal. *Ecology* 73:701–4.

Dinerstein, E., and K. Baragona. 2000. Saving cells is no way to save a species. *Washington Post,* May 28, p. B3.

Dinerstein, E., and G. F. McCracken. 1990. Endangered greater one-horned rhinoceros carry high levels of genetic variation. *Conservation Biology* 4:417–22.

Dinerstein, E., and J. N. Mehta. 1989. The clouded leopard in Nepal. *Oryx* 29:199–201.

Dinerstein, E., and L. Price. 1991. Demography and habitat use by greater one-horned rhinoceros in Nepal. *Journal of Wildlife Management* 55:401–11.

Dinerstein, E., and C. Wemmer. 1988. Fruits *Rhinoceros* eat: Dispersal of *Trewia nudiflora* (*Euphorbiaceae*) in lowland Nepal. *Ecology* 69:1768–74.

Dinerstein, E., and E. D. Wikramanayake. 1993. Beyond hot spots: How to prioritize investments in biodiversity conservation in the Indo-Pacific region. *Conservation Biology* 7:53–65.

Dinerstein, E., et al. 1997. *A Framework for Identifying High Priority Areas and Actions for the Conservation of Tigers in the Wild*. Washington, D.C.: World Wildlife Fund–United States, Wildlife Conservation Society, and National Fish and Wildlife Foundation.

Dinerstein, E., et al. 1999. Tigers as neighbors: Efforts to promote local guardianship of endangered species in lowland Nepal. In J. Seidensticker, S. Christie, and P. Jackson, eds., *Riding the Tiger: Tiger Conservation in Human-Dominated Landscapes*, pp. 316–33. Cambridge: Cambridge University Press.

Dinerstein, E., S. R. Shrestha, and H. Mishra. 1988. Adoption in greater one-horned rhinoceros *Rhinoceros unicornis*. *Journal of Mammalogy* 69:813–14.

Dinerstein, E., S. R. Shrestha, and H. Mishra. 1990. Capture, chemical immobilization, and radio-collar life for greater one-horned rhinoceros. *Wildlife Society Bulletin* 18:36–41.

Dinerstein, E., G. Zug, and J. Mitchell. 1987. Notes on the biology of *Melanochelys*. *Journal of the Bombay Natural History Society* 84:687–88.

Ecological Consulting. 1993. *Computer-Aided Mapping and Resource Inventory System*. Portland, Ore.: CAMRIS.

Eisenberg, J. F., and M. Lockhart. 1972. *An Ecological Reconnaissance of Wilpattu National Park, Ceylon*. Smithsonian Contributions to Zoology, no. 101. Washington, D.C.: Smithsonian Institution Press.

Eisenberg, J. F., and J. Seidensticker. 1976. Ungulates in southern Asia: A consideration of biomass estimates for selected habitats. *Biological Conservation* 10:298–308.

Ferraro, P. J. 2001. Global habitat protection: Limitations of development interventions and a role for conservation performance payments. *Conservation Biology* 15:990–1000.

Foose, T. J. 1982. Trophic strategies of ruminant versus nonruminant ungulates. Ph.D. diss., University of Chicago.

Foose, T. J. 1992. *IUCN SSC Global Captive Action Plan for Rhino*. Apple Valley, Minn.: International Union for Conservation of Nature and Natural Resources, Species Survival Commission, Conservation Breeding Specialists Group.

Foose, T. J. 1996. *International Studbook for the Sumatran Rhinoceros*. Cumberland, Ohio: International Rhino Foundation.

Foose, T. J., and R. Reece. 1996. *American Zoological Association Species Survival Plan: Master Plan for Rhinoceros*. Cumberland, Ohio: International Rhino Foundation.

Foose, T. J., and N. van Strien, eds. 1997. *Asian Rhinos — Status Survey and Conservation Action Plan*. Cambridge: International Union for Conservation of Nature and Natural Resources.

Foose, T. J., N. van Strien, and M. Khan. 1995. Letter to editor. *Conservation Biology* 9:977–84.

Frame, G. W. 1980. Black rhinoceros subpopulation on the Serengeti Plain, Tanzania. *African Journal of Ecology* 18:155–66.

Frankel, O. H., and M. E. Soulé. 1981. *Conservation and Evolution*. New York: Cambridge University Press.

Franklin, I. R. 1980. Evolutionary change in small populations. In M. E. Soulé and B. A. Wilcox, eds., *Conservation Biology: An Evolutionary-Ecological Perspective*, pp. 135–49. Sunderland, Mass.: Sinauer.

Freeman, G. H., and J. M. King. 1969. Relations amongst various linear measurements and weight for black rhinoceros in Kenya. *East African Wildlife Journal* 7:67–72.

Fuerst, P. A., and T. Maruyama. 1986. Considerations on the conservation of alleles and gene heterozygosity in small managed populations. *Zoo Biology* 5:171–79.

Garshelis, D. L. In press. Variation in Ursid life histories: Is there an outlier? In D. Lindburg and K. Baragona, eds., *The Giant Panda: Conservation Priorities for the Twenty-first Century*. Berkeley: University of California Press.

Gee, E. P. 1959. Report on survey of the rhinoceros area of Nepal. *Oryx* 5:59–85.

Global Environment Facility. 1999. *Ranging Patterns and Ecology of Elephants in Southern Sri Lanka*. Washington, D.C.: Global Environmental Facility; Colombo, Sri Lanka: Department of Wildlife Conservation.

Goddard, J. 1967. Home range behavior and recruitment rates of two black rhinoceros populations. *East African Wildlife Journal* 5:133–50.

Goddard, J. 1970. Age criteria and vital statistics of a black rhinoceros population. *East African Wildlife Journal* 8:105–21.

Goodman, D. E. 1987. The demography of chance extinction. In M. E. Soulé, ed., *Viable Populations for Conservation*, pp. 35–58. Cambridge: Cambridge University Press.

Graf, W., and L. Nichols Jr. 1966. The axis deer in Hawaii. *Journal of the Bombay Natural History Society* 63:629–734.

Groves, C. P. 1993. Testing rhinoceros subspecies by multivariate analysis. In O. A. Ryder, ed., *Rhinoceros Biology and Conservation*, pp. 92–100. San Diego, Calif.: Zoological Society of San Diego.

Hanley, T. A., and R. D. Taber. 1980. Selective plant species inhibition by elk and deer in three conifer communities in western Washington. *Forest Science* 26:97–107.

Harris, H., and D. A. Hopkinson. 1978. *Handbook of Enzyme Electrophoresis in Human Genetics*. New York: American Elsevier.

Hedrick, P. W., et al. 1986. Protein variation, fitness, and captive propagation. *Zoo Biology* 5:91–99.

Heissig, K. 1989. The Rhinocerotidae. In D. R. Prothero and R. M. Schoch, eds., *The Evolution of Perissodactyls*, pp. 399–418. New York: Oxford University Press.

Hiley, P. G. 1977. The thermoregulatory response of the rhinoceros (*Diceros bicornis* and *Ceratotherium simum*) and the zebra to diurnal temperature change. *East Africa Wildlife Journal* 13:337.

Hillman-Smith, K., et al. 1986. Age estimation of the white rhinoceros (*Ceratotherium simum*). *Journal of the Zoological Society of London* 210:355–77.

Hitchins, P. M. 1978. Age determination of the black rhinoceros in Zululand. *South African Journal of Wildlife Research* 8:71–80.

Hoogerwerf, A. 1970. *Udjung Kulon, the Land of the Last Javan Rhinoceros.* Leiden: Brill.

Howe, H. F. 1984. Constraints on the evolution of mutualisms. *American Naturalist* 123:764–77.

Howe, H. F. 1985. Gomphothere fruits: A critique. *American Naturalist* 125: 853–65.

Hutchins, M., R. Wiese, and K. Willis. 1995. Letter to editor. *Conservation Biology* 9:977.

Ivlev, V. S. 1961. *Experimental Ecology of the Feeding Fishes.* New Haven, Conn.: Yale University Press.

Janis, C. 1976. The evolutionary strategy of the *Equidae* and the origins of rumen and caecal digestion. *Evolution* 30:757–76.

Janzen, D. H. 1981a. *Enterolobium cyclocarpum* seed passage rate and survival in horses, Costa Rican Pleistocene seed dispersal agents. *Ecology* 62:593–601.

Janzen, D. H. 1981b. Guanacaste tree seed-swallowing by Costa Rican range horses. *Ecology* 62:587–92.

Janzen, D. H. 1981c. Digestive seed predation by a Costa Rican Baird's tapir. *Biotropica* 13:59–63.

Janzen, D. H. 1982a. Differential seed survival and passage rates in cows and horses, surrogate Pleistocene dispersal agents. *Oikos* 38:150–56.

Janzen, D. H. 1982b. Seeds in tapir dung in Santa Rosa National Park, Costa Rica. *Brenesia* 19:129–35.

Janzen, D. H. 1984. Dispersal of small seeds by big herbivores: Foliage is the fruit. *American Naturalist* 123:338–53.

Janzen, D. H. 1986. Chihuahuan Desert Nopaleras: Defaunated big mammal vegetation. *Annual Review of Ecology and Systematics* 17:595–636.

Janzen, D. H., and P. Martin. 1982. Neotropical anachronisms: What the gomphotheres ate. *Science* 215:19–27.

Jnawali, S. R. 1986. Diet analysis of greater one-horned rhinoceros (*Rhinoceros unicornis*) by fecal analysis. Master's thesis, Tribhuvan University, Kathmandu, Nepal.

Jnawali, S. R. 1995. Population ecology of greater one-horned rhinoceros (*Rhinoceros unicornis*) with particular emphasis on habitat preference, food ecology, and ranging behavior of a reintroduced population in Royal Bardia National Park, Nepal. Ph.D. diss., Agricultural University of Norway, Oslo.

Joshi, A. R., E. Dinerstein, and D. Smith. 2002. The Terai Arc: Managing tigers and other wildlife as metapopulations. In E. Wikramanayake et al., eds., *Terrestrial Ecoregions of the Indo-Pacific: A Conservation Assessment,* pp. 178–81. Washington, D.C.: Island Press.

Kaziranga under water. 1988. *Himal Magazine.* November–December, p. 33.

Keiter, R. B. 1995. Preserving Nepal's national parks: Law and conservation in the developing world. *Ecology Law Quarterly* 22:591–675.

Kenny, J. S., D. Smith, A. M. Starfield, and C. McDougal. 1995. The long-term effects of tiger poaching on population viability. *Conservation Biology* 9:1127–33.

Bin Khan, M. K. 1989. *Asian Rhinos: An Action Plan for Their Conservation.* Gland, Switzerland: International Union for Conservation of Nature.

Lande, R. 1988. Genetics and demography in biological conservation. *Science* 241: 1455–60.

Lande, R., and G. F. Barrowclough. 1987. Effective population size, genetic variation, and their use in population management. In M. E. Soulé, ed., *Viable Populations for Conservation,* pp. 87–123. Cambridge: Cambridge University Press.

Lang, E. M. 1961. Beobachtungen an Indischen Panzernashorn (*Rhinoceros unicornis*). *Liepzig Zoological Gartens* 25:1–39.

Langman, V. A. 1985. Heat balance in the black rhinoceros (*Diceros bicornis*). *National Geographic Research Reports* 21:251–54.

Laurie, W. A. 1978. The ecology of the greater one-horned rhinoceros. Ph.D. diss., Cambridge University.

Laurie, W. A. 1982. Behavioural ecology of greater one-horned rhinoceros (*Rhinoceros unicornis*). *Journal of the Zoological Society of London* 196:307–41.

Laurie, W. A., E. M. Lang, and C. P. Groves. 1983. *Rhinoceros unicornis. Mammalian Species* 211:1–6.

Leader-Williams, N. 1993. Theory and pragmatism in the conservation of rhinos. In O. A. Ryder, ed., *Rhinoceros Biology and Conservation,* pp. 69–81. San Diego, Calif.: Zoological Society of San Diego.

Lehmkuhl, J. 1989. The ecology of a south Asian tallgrass community. Ph.D. diss., University of Washington.

Levene, H. 1949. On a matching problem arising in genetics. *Annals of Mathematics and Statistics* 20:91–94.

Lindeque, M., and K. P. Erb. 1995. Research on the effects of temporary horn removal on black rhinos in Namibia. *Pachyderm* 20:27–30.

Lucas, S. G., and J. Sobus. 1989. The systematics of Indricotheres. In D. R. Prothero and R. M. Schoch, eds., *The Evolution of Perissodactyl,* pp. 358–79. New York: Oxford University Press.

MacKinnon, K., H. Mishra, and J. Mott. 1999. Reconciling the needs of conservation and local communities: Global environment facility support for tiger conservation in India. In J. Seidensticker, S. Christie, and P. Jackson, eds., *Riding the Tiger: Tiger Conservation in Human-Dominated Landscapes,* pp. 307–15. Cambridge: Cambridge University Press.

Martin, E. B., and C. B. Martin. 1987. Combating the illegal trade in rhinoceros products. *Oryx* 21:143–48.

Martin, E. B., L. Vigne, and C. Allan. 1997. *On a Knife's Edge: The Rhinoceros Horn Trade in Yemen.* Cambridge: Traffic International.

Maskey, T. M. 1979. *Royal Chitwan National Park: Report on Gharial.* Kathmandu, Nepal: Department of National Parks and Wildlife Conservation.

McCracken, G. F. 1984. Communal nursing in Mexican free-tailed bat maternity colonies. *Science* 223:1090–91.

McCracken, G. F., and G. S. Wilkinson. 1988. Allozyme techniques and kinship assessment in bats. In T. H. Kunz, ed., *Ecological and Behavioral Methods for the Study of Bats,* pp. 141–55. Washington, D.C.: Smithsonian Institution Press.

McKay, G. M. 1973. *Behavior and Ecology of the Asiatic Elephant in Southeastern Ceylon.* Smithsonian Contributions to Zoology, no. 125. Washington, D.C.: Smithsonian Institution Press.

McNaughton, S. J. 1984. Grazing lawns: Animals in herds, plant form, and coevolution. *American Naturalist* 124:863–86.

McShea, W. J., and J. H. Rappole. 2000. Managing the abundance and diversity of breeding bird populations through manipulation of deer populations. *Conservation Biology* 14:1161–71.

Meijaard, E. 1996. The Sumatran rhinoceros in Kalimantan, Indonesia: Its possible distribution and conservation prospects. *Pachyderm* 21:15–23.

Menon, V. 1996. *Under Siege: Poaching and Protection of Greater One-Horned Rhinoceroses in India.* Cambridge: Traffic International.

Merenlender, A. M., et al. 1989. Allozyme variation and differentiation in African and Indian rhinoceroses. *Journal of Heredity* 80:377–82.

Milton, J. P., and G. A. Binney. 1980. *Ecology Planning in the Nepalese Terai: A Report on Resolving Resource Conflicts Between Wildlife Conservation and Agricultural Land-use in Padampur Panchayat.* Washington, D.C.: International Center for Environmental Renewal.

Mishra, H. B. 1982. The ecology and behavior of chital (*Axis axis*) in the Royal Chitwan National Park, Nepal. Ph.D. diss., University of Edinburgh.

Mishra, H. R., and E. Dinerstein. 1987. New zip codes for resident rhinos in Nepal. *Smithsonian Magazine* 18:66–73.

Mishra, H. R., and M. Jeffries. 1991. *Royal Chitwan National Park: Wildlife Heritage of Nepal.* Seattle: The Mountaineers.

Mueller-Dombois, D. 1972. Crown distortion and elephant distribution in the woody vegetation of Ruhuna National Park, Ceylon. *Ecology* 53:208–26.

National Planning Commission Secretariat. 1994. *Population of Nepal: Municipalities 1991 Population Census.* Ramshah Path, Kathmandu: Central Bureau of Statistics.

Nei, M. 1987. *Molecular Evolutionary Genetics.* New York: Columbia University Press.

Nei, M., T. Maruyama, and R. Chakraborty. 1975. The bottleneck effect and genetic variability in populations. *Evolution* 29:1–10.

Noss, R. F., et al. 1999. Core areas: Where nature reigns. In M. E. Soulé and J. Terborgh, eds., *Continental Conservation: Scientific Foundations of Regional Reserve Networks,* pp. 99–129. Washington, D.C.: Island Press.

Noss, R. F., M. A. O'Connell, and D. D. Murphy. 1997. *The Science of Conservation Planning: Habitat Conservation Under the Endangered Species Act.* Washington, D.C.: Island Press.

Oates, J. F. 1999. *Myth and Reality in the Rain Forest: How Conservation Strategies Are Failing in West Africa.* Berkeley: University of California Press.

O'Brien, S. J., and J. F. Evermann. 1988. Interactive influence of infectious disease and genetic diversity in natural populations. *Trends in Ecology and Evolution* 3:254–59.

O'Brien, S. J., et al. 1985. Genetic basis for species vulnerability in the cheetah. *Science* 227:1428–34.

O'Brien, S. J., et al. 1987. East African cheetahs: Evidence for two population bottlenecks? *Proceedings of the National Academy of Sciences of the United States of America* 84:508–11.

Oldfield, H. A. 1880. *Sketches from Nepal, Historical and Descriptive.* London: Allen.

Oliver, R. L. W. 1980. The pygmy hog: The biology and conservation of the pigmy hog (*Susporcula salvanius*), and the hispid hare (*Capralagus hispidus*). Special Scientific Report 1. *Jersey Wildife Preservation Trust* 1:1–80.

Olivier, R. 1978. Ecology of the Asiatic elephant (*Elephas maximus*). Ph.D. diss., Cambridge University.

Olson, D. M., and E. Dinerstein. 1998. The Global 200: A representation approach to conserving the earth's most biologically valuable ecoregions. *Conservation Biology* 12:502–15.

Osborn, H. F. 1923. The extinct giant rhinoceros *Baluchitherium* of western and Central Asia. *Natural History* 23:208–28.

Owen-Smith, N. 1973. The behavioral ecology of the white rhinoceros. Ph.D. diss., University of Wisconsin.

Owen-Smith, N. 1981. The white rhino overpopulation problem and a proposed solution. In P. A. Jewell and S. Holt, eds., *Problems in Management of Locally Abundant Wild Mammals,* pp. 129–50. New York: Academic Press.

Owen-Smith, N. 1987. Pleistocene extinctions: The pivotal role of megaherbivores. *Paleobiology* 13:351.

Owen-Smith, N. 1988. *Megaherbivores: The Influence of Very Large Body Size on Ecology.* Cambridge Studies in Ecology. Cambridge: Cambridge University Press.

Pellinck, E., and B. N. Upreti. 1972. *A Census of Rhinoceros in Chitwan National Park and Tamaspur Forest, Nepal.* Kathmandu: Nepal Food and Agriculture Organization, United Nations Development Program, and National Parks and Wildlife Conservation Project.

Pemberton, J. M., and R. H. Smith. 1985. Lack of biochemical polymorphism in British fallow deer. *Heredity* 55:199–207.

Peters, R. H. 1983. *The Ecological Implications of Body Size.* Cambridge: Cambridge University Press.

Pimm, S. L., J. L. Gittleman, G. F. McCracken, and M. F. Gilpin. 1989. Plausible alternatives to bottlenecks to explain reduced genetic diversity. *TREE* 4:176–78.

Population Reference Bureau. 1995. *World Population Data Sheet: Demographic Data and Estimates for the Countries and Regions of the World.* Washington, D.C.: Population Reference Bureau.

Prothero, D. R. 1993. Fifty million years of rhinoceros evolution. In O. A. Ryder, ed., *Rhinoceros Biology and Conservation,* pp. 81–87. San Diego, Calif.: Zoological Society of San Diego.

Prothero, D. R., and R. M. Schoch, eds. 1989. *The Evolution of Perissodactyls.* New York: Oxford University Press.

Prothero, D. R., C. Guerin, and E. Manning. 1989. The history of Rhinocerotidae. In D. R. Prothero and R. M. Schoch, eds., *The Evolution of Perissodactyls,* pp. 322–40. New York: Oxford University Press.

Rabinowitz, A. 1995. Letter to editor. *Conservation Biology* 9:482–88.

Ralls, K. 1976. Mammals in which females are larger than males. *Quarterly Review of Biology* 51:245–76.

Ralls, K., and P. Harvey. 1985. Geographic variation in size and sexual dimorphism in North American weasels. *Biological Journal of the Linnaean Society* 25: 119–67.

Reynolds, R. J. 1960. Asian rhinos in captivity. *International Zoological Yearbook* 2:17–42.

SAS. 1989. *SAS/STIT Users' Guide.* 4th ed. Version 6, vol. 2. Cary, N.C.: SAS Institute.

Schaller, G. B. 1967. *The Deer and the Tiger.* Chicago: University of Chicago Press.

Schaller, G. B. 1988. *Stones of Silence: Journeys in the Himalaya.* Chicago: University of Chicago Press.

Schmidt-Nielsen, K. 1984. *Scaling: Why Is Animal Size So Important?* Cambridge: Cambridge University Press.

Schonewald-Cox, C. M., S. M. Chamber, B. MacBryde, and L. Thomas. 1983. *Genetics and Conservation: A Reference for Managing Wild Animal and Plant Populations.* London: Benjamin/Cummings.

Seidensticker, J. 1976. Ungulate populations in Chitwan Valley, Nepal. *Biological Conservation* 10:183–210.

Seidensticker, J., S. Christie, and P. Jackson, eds. 1999. *Riding the Tiger: Tiger Conservation in Human-Dominated Landscapes.* Cambridge: Cambridge University Press.

Selander, R. K., M. H. Smith, S. Y. Yang, W. E. Johnson, and J. B. Gentry. 1971. Biochemical polymorphism and systematics in the genus *Peromyscus.* I. Variation in the old field mouse. *Studies in Genetics* 6:49–90.

Seshadri, B. K. 1969. *The Twilight of India's Wildlife.* London: Baker.

Sharma, U. R. 1991. Park–people interactions in Royal Chitwan National Park, Nepal. Ph.D. diss., University of Arizona.

Sinha, S. P., and V. B. Sawarkar. 1993. Management of the reintroduced greater one-horned rhinoceros (*Rhinoceros unicornis*) in Dudhwa National Park, Uttar Pradesh, India. In O. A. Ryder, ed., *Rhinoceros Biology and Conservation,* pp. 218–27. San Diego, Calif.: Zoological Society of San Diego.

Smith, J. L. D., C. McDougal, and M. E. Sunquist. 1987. Female and tenure system in tigers. In R. L. Tilson and U. S. Seal, eds., *Tigers of the World: The Biology, Biopolitics, Management, and Conservation of an Endangered Species,* pp. 97–109. Park Ridge, N.J.: Noyes.

Smythies, E. A. 1942. *Big Game Hunting in Nepal.* London: Thacker, Spink.

Soulé, M. E. 1976. Allozyme variation: Its determinants in space and time. In F. J. Ayala, ed., *Molecular Evolution,* pp. 60–77. Sunderland, Mass.: Sinauer.

Soulé, M. E. 1996. The end of evolution? *World Conservation,* April, pp. 24–25.

Spillett, J. J. 1967a. A report on wildlife surveys in north India and southern Nepal: The large mammals of the Keoladeo Ghana Sanctuary, Rajasthan. *Journal of the Bombay Natural History Society* 63:602–7.

Spillett, J. J. 1967b. A report on wildlife surveys in north India and southern Nepal: The Jaldapara Wild Life Sanctuary, West Bengal. *Journal of the Bombay Natural History Society* 63:534–56.

Spillett, J. J., and K. M. Tamang. 1967. Wildlife conservation in Nepal. *Journal of the Bombay Natural History Society* 63:557–71.

Spitsin, V., V. P. Romonov, N. S. Popov, and E. N. Simirnov. 1987. The Siberian tiger (*Panthera tigris altaica* — Temminck 1844) in the USSR: Status in the wild and in captivity. In R. L. Tilson and U. S. Seal, eds., *Tigers of the World: The Biology, Biopolitics, Management, and Conservation of an Endangered Species,* pp. 64–70. Park Ridge, N.J.: Noyes.

Stracey, P. D. 1957. On the status of the great Indian rhinoceros (*R. unicornis*) in Nepal. *Journal of the Bombay Natural History Society* 54:763–66.

Sukumar, R. 1999. *The Asian Elephant: Ecology and Management.* Cambridge: Cambridge University Press.

Sumardja, E. 1995. Letter to editor. *Conservation Biology* 9:977–84.

Sunquist, M. E. 1981. *The Social Organization of Tigers (Panthera tigris) in Royal Chitwan National Park, Nepal.* Smithsonian Contributions to Zoology, no. 336. Washington, D.C.: Smithsonian Institution Press.

Sunquist, M. E., U. K. Karanth, and F. Sunquist. 1999. Ecology, behaviour, and resilience of the tiger and its conservation needs. In J. Seidensticker, S. Christie, and P. Jackson, eds., *Riding the Tiger: Tiger Conservation in Human-Dominated Landscapes,* pp. 5–19. Cambridge: Cambridge University Press.

Taylor, C. R. 1970. Dehydration and heat: Effect on temperature regulation of East African ungulates. *American Journal of Physiology* 219:1136–39.

Terborgh, J. 1999. *Requiem for Nature.* Washington, D.C.: Island Press.

Terborgh, J., and B. Winter. 1980. Some causes of extinction. In M. E. Soulé and B. A. Wilcox, eds., *Conservation Biology: An Evolutionary-Ecological Perspective,* pp. 119–33. Sunderland, Mass.: Sinauer.

Thompson, J. 1982. *Interaction and Coevolution.* New York: Wiley.

Wall, W. P. 1989. The phylogenetic history and adaptive radiation of the Amynodontidae. In D. R. Prothero and R. M. Schoch, eds., *The Evolution of Perissodactyls,* pp. 341–55. New York: Oxford University Press.

Wehausen, J. D. 1999. Rapid extinction of mountain sheep populations revisited. *Conservation Biology* 13:378–84.

Wells, M. 1993. Neglect of biological riches: The economies of nature tourism in Nepal. *Biodiversity and Conservation* 2:445–64.

Wells, M., et al. 1999. *Investing in Biodiversity: A Review of Indonesia's Integrated Conservation and Development Projects.* Washington, D.C.: World Bank.

Wemmer, C., R. Simons, and H. R. Mishra. 1983. Case study of a cooperative international conservation program: The tiger ecology project. Paper presented at the Bombay Natural History Society Centenary Seminar, Bombay.

Western, D., and R. M. Wright. 1994. *Natural Connections: Perspectives in Community-based Conservation.* Washington, D.C.: Island Press.

Wheelwright, N. T., and G. H. Orians. 1982. Seed dispersal by animals: Contrasts with pollen dispersal, problems of terminology, and constraints on coevolution. *American Naturalist* 819:402–13.

Widdowson, E. M. 1976. The response of the sexes to nutritional stress. *Proceedings of the Nutrition Society* 35:175–80.

Wikramanayake, E., et al. 1998a. An ecology-based method of defining priorities for large mammal conservation: The tiger as case study. *Conservation Biology* 12: 865–78.

Wikramanayake, E. 1998b. *A Biodiversity Assessment and Gap Analysis of the Himalayas.* Washington, D.C.: Conservation Science Program, World Wildlife Fund–United States, and United Nations Development Program.

Wikramanayake, E., et al. 1999. Where can tigers live in the future? A framework for identifying high-priority areas for the conservation of tigers in the wild. In J. Seidensticker, S. Christie, and P. Jackson, eds. *Riding the Tiger: Tiger Conservation in Human-Dominated Landscapes,* pp. 255–73. Cambridge: Cambridge University Press.

Wikramanayake, E., et al., eds. 2002. *Terrestrial Ecoregions of the Indo-Pacific: A Conservation Assessment.* Washington, D.C.: Island Press.

Wilkinson, L. 1989. *The System for Statistics.* Evanston, Ill.: Systat.

Wolanski, N. 1979. The adult. In D. B. Jellicliffe and E. F. Patrice Jellicliffe, eds., *Human Nutrition: A Comprehensive Treatise.* Vol. 2, *Nutrition and Growth,* pp. 254–72. New York: Plenum.

Woods, M. 1984. *Rhinos.* Vol. 1 of *Zoobooks.* San Diego, Calif.: Wildlife Education.

Wooten, M. D., and M. H. Smith. 1985. Large mammals are genetically less variable? *Evolution* 39:210–12.

World Wildlife Fund. 2002. *Securing a Future for Asia's Wild Rhinos and Elephants: WWF's Asian Rhino and Elephant Action Strategy.* Washington, D.C.: World Wildlife Fund.

Yonzon, P. B. 1994. *Count Rhino '94.* World Wildlife Fund Nepal Program Report to the Department of National Parks and Wildlife Conservation, King Mahendra Trust for Nature Conservation and Wildlife Fund, Katmandu. Washington, D.C. World Wildlife Fund.

Zar, J. H. 1984. *Biostatistical Analysis.* 2d ed. Englewood Cliffs, N.J.: Prentice-Hall.

Zingeser, M. R., and C. R. Phoenix. 1978. Metric characteristics of the canine dental complex in prenatally androgenized female rhesus monkeys. *American Journal of Physical Anthropology* 49:187–92.

Zug, G. R., and J. C. Mitchell. 1995. Amphibians and reptiles of Royal Chitwan National Park, Nepal. *Asiatic Herpetological Research* 6:172–80.

INDEX

Italic page numbers indicate photographs, and boldface page numbers indicate material in charts, maps, and tables.